Series on Technology
and Social Priorities

NATIONAL ACADEMY
OF ENGINEERING

Technology in Services:
Policies for Growth, Trade, and Employment

Bruce R. Guile and James Brian Quinn
Editors

NATIONAL ACADEMY PRESS
Washington, D.C. 1988

National Academy Press ● 2101 Constitution Avenue, NW ● Washington, DC 20418

NOTICE: The National Academy of Engineering was established in 1964, under the charter of the National Academy of Sciences, as a parallel organization for outstanding engineers. It is autonomous in its administration and in the selection of its members, sharing with the National Academy of Sciences the responsibility for advising the federal government. The National Academy of Engineering also sponsors engineering programs aimed at meeting national needs, encourages education and research, and recognizes the superior achievement of engineers. Dr. Robert M. White is president of the National Academy of Engineering.

Funds for the National Academy of Engineering's Symposium Series on Technology and Social Priorities were provided by the Andrew W. Mellon Foundation, Carnegie Corporation of New York, and the Academy's Technology Agenda Program. This publication has been reviewed by a group other than the authors according to procedures approved by a Report Review Committee. The views expressed in this volume are those of the authors and are not presented as the views of the Mellon Foundation, Carnegie Corporation, or the National Academy of Engineering.

Library of Congress Cataloging-in-Publication Data

Technology in services: policies for growth, trade, and employment /
Bruce R. Guile and James Brian Quinn, editors.

    p. cm.—(Series on technology and social priorities)
    At head of title: National Academy of Engineering.
    Includes material presented at an NAE symposium entitled
"Technology in Services: the Next Economy" held in Washington, DC,
on January 28 and 29, 1988.
    Bibliography: p.
    Includes index.
    ISBN 0-309-03895-2.—ISBN 0-309-03887-1 (pbk.)
    1. Service industries—United States—Technological innovations—
Congresses. I. Guile, Bruce R. II. Quinn, James Brian, 1928–
III. National Academy of Engineering.
HD9981.5.T43 1988
338.4'561'000973—dc19   88-37920
                                            CIP

Printed in the United States of America

Preface

It is common to structure the debate about U.S. competitiveness in a global economy principally in terms of the productivity growth and trade performance of U.S. manufacturing industries. Although it is true that manufacturing industries are a crucial element of our national economy, it is important to recognize that the U.S. economy consists of interdependent productive systems, including manufacturing, services, agriculture, mining, and natural resources. Although each of these sectors exhibits considerable internal diversity, each sector is dependent on all other sectors for the level of its productivity. Services industries, in particular, play a central role in this system. Services now account for the largest part of U.S. gross national product and of employment, and industries such as communications, finance, transportation, and medicine are technologically dynamic and absolutely central elements of the U.S. and global economies. They are the heart and circulatory system of the current and next global economy.

This volume and its companion volume, *Managing Innovation: Cases from the Services Industries*, explore the role, importance, and technological dynamism of services industries. *Managing Innovation* focuses on the application of technologies in services businesses. This volume addresses the way in which technology applied in services business has changed and continues to affect the structure of production. In particular, the volume focuses on economic structural change, issues of services productivity, and trade in services in a way that should be helpful to policymakers. Perhaps the most important lesson that the volume offers is that policymakers concerned with the performance of the U.S. economy need to accept the central role of services industries.

Much of the material in this volume was presented at an NAE symposium entitled "Technology in Services: The Next Economy" held in Washington, D.C., on January 28 and 29, 1988. I would like to thank James Brian Quinn, who chaired the activity on technology in services, and Bruce R. Guile, the principal staff officer for the project, for their efforts in organizing the symposium and moving quickly to get this material published. Also, on behalf of the National Academy of Engineering, I would like to thank the symposium advisory committee (listed on p. 235) and the authors who participated in the symposium for their time and energy. Special thanks are due to Penny Cushman Paquette, research associate at the Amos Tuck School of Business at Dartmouth, and to a number of individuals who worked on the project or publication including Jesse H. Ausubel, H. Dale Langford, Hedy E. Sladovich, Michele M. Rivard, and Bette R. Janson, National Academy of Engineering, and Sally Fields, National Academy Press.

ROBERT M. WHITE
President
National Academy of Engineering

Contents

SERVICES TRADE AND REGULATION

SERVICES AND POLICY DIRECTIONS

Technology
in Services

Introduction to Services Industries Policy Issues

BRUCE R. GUILE

Services industries allow the management and financing of a nation's productive system and are the way an economy organizes itself to meet essential needs such as health care, transportation, communications, education, and the distribution of goods. Services industries are core economic functions in any economy and are the dominant economic activity in industrialized nations, accounting for the majority of both jobs and national output. Also, modern services industries are technologically dynamic—actively engaged in the development and application of virtually all advanced technologies. The companion volume to this book, *Managing Innovation: Cases from the Services Industries* (Guile and Quinn, 1988), samples services industries to illustrate the contribution of technological advance in services such as stock exchanges, banking, cellular telephone services, automotive repair, engineering design, and express package transportation. This is not to propose that services are somehow more important than manufacturing, agriculture, or natural resources industries but rather to initiate the argument that there is a technological/ecological principle at work that renders each sector dependent on all other sectors for the level of its productivity. This volume explores these relationships and the growing reliance on services industries for the management and control of an ever more complex formal economy. In exploring the importance and technological dynamism of services industries, the volume joins, indirectly, an active policy debate about U.S. deindustrialization.

1

SERVICES AND THE DEINDUSTRIALIZATION DEBATE

In recent years, large U.S. merchandise trade deficits, the shrinking of several traditionally important U.S. manufacturing industries, and the growth in services industries employment have spawned an active debate about U.S. "deindustrialization" (see Bluestone and Harrison, 1982, 1986; Cohen and Zysman, 1987; Lawrence, 1984). The concept of deindustrialization brings together a wide array of concerns. Is the U.S. level of investment in manufacturing plant and equipment sufficient? Are U.S.-based manufacturers keeping up with foreign competitors in terms of production technology? Does the loss of certain manufacturing jobs to foreign-based production have irreversible consequences for the level and distribution of U.S. wages and household incomes? These are indeed critical national issues; manufacturing performance is an important element of economic performance in any large industrialized economy.

The problem arises in posing the issue of deindustrialization as one of manufacturing versus services. The growing importance of services industries in the U.S. economy is often mistakenly cited as evidence of deindustrialization, as evidence of a crisis in U.S. economic performance. In reality, goods and services industries are wholly interdependent and equally important elements of the U.S. economy. Additionally, the shift to "a services economy" has not been a rapid, dramatic change but a steady trend, seemingly consistent with a sound, balanced, industrial economy. That conclusion arises from considering long-term trends in output and employment.

With regard to output, the most commonly used data series is the Gross Product Originating (GPO) series published by the Department of Commerce Bureau of Economic Analysis. Table 1 shows the level of output and share of output in selected major sectors of the U.S. economy between 1948 and 1987. As Table 1 illustrates, durable and nondurable goods manufacturing (and retail trade) have grown in absolute terms but are virtually unchanged in magnitude relative to the size of the whole economy. Several sectors have diminished in relative size. Agriculture, forestry, and fisheries; mining; transportation; construction; and government account for significantly smaller portions of national output than they did in 1948. The relative decrease in those sectors has been accompanied by significant increases in the share of output contributed by communications; wholesale trade; finance, insurance, and real estate; and services. The services grouping includes a wide variety of services including health, business, legal, amusement, and hotel services. In summary, although some services industries have grown substantially in importance since the late 1940s, the data do not show a decline in the relative contribution of manufacturing output to the U.S. national economy (for further discussion, see Cremeans, 1985; Lawrence, 1984; Moody, 1985). Although the validity of the GPO series has been challenged for not accurately

TABLE 1 Gross Product Originating in Selected Industries for Selected Years, 1948–1987 (billions of 1982 dollars and percent share of Gross National Product)

Sector	1948	1957	1969	1979	1984	1987
Real gross national product	1,108.7	1,551.1	2,423.3	3,192.4	3,501.4	3,847.0
	100.0%	100.0%	100.0%	100.0%	100.0%	100.0%
Agriculture, forestry, fisheries	61.3	65.9	65.3	76.1	82.2	96.1
	5.5%	4.2%	2.7%	2.4%	2.3%	2.5%
Mining	72.4	96.2	128.9	130.0	133.0	117.5
	6.5%	6.2%	5.3%	4.1%	3.8%	3.1%
Construction	90.0	142.4	183.6	173.5	159.2	175.8
	8.1%	9.2%	7.6%	5.4%	4.5%	4.6%
Durable goods manufacturing	145.0	208.7	334.1	423.5	466.8	525.2
	13.1%	13.5%	13.8%	13.3%	13.3%	13.7%
Nondurable goods manufacturing	93.5	123.8	202.6	273.5	291.1	314.3
	8.4%	8.0%	8.4%	8.6%	8.3%	8.2%
Transportation	76.5	75.0	104.2	137.7	123.7	136.0
	6.9%	4.8%	4.3%	4.3%	3.5%	3.5%
Communications	9.0	16.6	37.4	72.5	92.9	107.6
	0.8%	1.1%	1.5%	2.3%	2.7%	2.8%
Electricity, gas, sanitation	13.2	28.3	58.6	83.3	103.8	105.9
	1.2%	1.8%	2.4%	2.6%	3.0%	2.8%
Wholesale trade	55.8	80.8	149.0	217.3	250.6	291.7
	5.0%	5.2%	6.1%	6.8%	7.2%	7.6%
Retail trade	106.1	144.3	212.7	294.4	328.3	368.3
	9.6%	9.3%	8.8%	9.2%	9.4%	9.6%
Finance, insurance, real estate	107.7	178.3	314.0	459.2	506.6	559.4
	9.7%	11.5%	13.0%	14.4%	14.5%	14.5%
Services[a]	128.9	168.6	287.8	429.8	514.0	610.8
	11.6%	10.9%	11.9%	13.5%	14.7%	15.9%
Government	155.5	229.2	340.2	376.2	392.1	415.7
	14.0%	14.8%	14.0%	11.8%	11.2%	10.8%

[a]Health and business services account for more than half the output of this sector in 1987. Many other services activities are part of this category, including legal services, motion pictures, automobile repair, and amusement and recreation services.

SOURCE: U.S. Department of Commerce, Bureau of Economic Analysis.

reflecting recent events in the U.S. manufacturing sector (Mishel, 1988), there is no better long-term data series available for examining U.S. economic output.

The story is slightly different with regard to employment by sector. Employment in the U.S. manufacturing sector has decreased as a percentage of total U.S. employment, from 34 percent in 1950 to 19 percent in 1986. This has occurred, however, as a result of growth in other sectors. The absolute level of the manufacturing work force has fluctuated quite dramatically with economic conditions, but it shows no clear trend similar to that in manufacturing's share of total employment. The fact that the manufacturing sector's share of a growing gross national product is steady while its contribution to actual employment remains constant is consistent with long-term trends of increasing labor productivity in manufacturing.

The diminishing share of manufacturing in total employment over the last 35 years reflects increases in employment share across a wide variety of sectors. There have been small employment share increases in wholesale trade and small decreases in agriculture, mining, transportation, and utilities. There have been substantial increases in retail trade, business services, health services, and government, mostly state and local government (which includes education). Services-producing industries have been responsible for virtually all of the job growth in the United States since 1972, and in 1986 they employed over 70 percent of the work force.

The common belief that this shift in employment shares reflects a shift from high-wage factory work to low-wage retail and food services work is belied by a simple examination of the areas in which employment has grown. Although it is true that there has been significant employment growth in retail trade, there has also been substantial growth in business services, health services, and government. Whereas compensation in the retail industry is quite low relative to all-industry averages (64 percent of the All Domestic Industry Average in 1983) compensation is much higher in business services (87 percent), and health services and government are very close to the mean (97 and 104 percent, respectively) (Moody, 1985). In contrast, manufacturing was 118 percent in 1983. A recent report from the Committee on Science, Engineering, and Public Policy of the National Academy of Sciences/National Academy of Engineering/Institute of Medicine described the changes this way (Cyert and Mowery, 1987, p. 109):

[T]he characteristic form of structural change within this economy does not involve a large net outflow of labor from manufacturing into nonmanufacturing employment; rather, it reflects more rapid employment growth in industries in which average wage rates currently are lower than in manufacturing. At the same time, however, the occupational structure of the U.S. economy has shifted in an opposite direction, with faster growth in higher-skill, higher-wage occupations. . . . Partly for this reason the gap in average wages between manufacturing and rapidly growing sectors such as business services (which include computer services and consulting) has been shrinking over the past decade. . . . Many declining manufacturing industries—for example, textile, apparel, and leather products—now pay wages that are low in comparison with those paid by much of the services sector.

Although the relative shift in employment toward nonmanufacturing industries has probably been responsible for some slowing of the rate of growth in real earnings in the United States, the effects of the shift must be interpreted carefully. If the United States is becoming a nation of hamburger stands, it is also becoming a nation of management consultants, doctors, software designers, and international bankers.

In summary, the policy directions suggested by the changing structure of the U.S. economy are ambiguous. On the one hand, it is quite clear that manufacturing productivity growth is crucial to U.S. economic welfare. Manufacturing is critical for obvious national security reasons and also because manufacturing activities have traditionally represented the primary opportunity to add value in a manner that was easily exported.

On the other hand, although it is true that the relative role of manufacturing and manufacturing employment in the U.S. economy is changing, the changes are more complex than those implied by the phrase "deindustrialization." Manufacturing activities have not disappeared from the United States, nor does it seem likely that they will disappear in the near future. The composition of employment has shifted toward nonmanufacturing industries but not exclusively into employment that pays poorly. The chapters in this volume, while not directed at questions about deindustrialization, do offer some important insights about the roles of both services and manufacturing industries in the operation and productivity of the global economy.

THE ROLE AND IMPORTANCE OF SERVICES

Although many individuals think of a mass-producing manufacturer as the prototypical large company, U.S. economic history might lead one to a different conclusion. Railroad companies—services companies—emerged during the mid-1800s as the first modern corporations. The business historian Alfred Chandler (1977, p. 120) described railroads as follows:

They were the first to require a large number of salaried managers; the first to have a central office operated by middle managers and commanded by top managers who reported to a board of directors. They were the first American business enterprise to build a large internal organizational structure with carefully defined lines of responsibility, authority, and field units.

It was not until the early 1900s that the managerial and financial systems developed for managing railroads made their way into production operations and, indeed, it was not until 1913 with Ford's innovation of the moving assembly line that mass manufacturing, as well as the size and structure of manufacturing firms, took off. No single services industry today is as dominant as the railroads were in their time, but communications firms, banks, airlines, multiple-site retailers, and energy utilities are large, sophisticated corporations with as much to contribute to national economic well-being as large manufacturing firms.

In the opening chapter of this volume, James Brian Quinn examines the role and character of services industries. The chapter begins by addressing some common misperceptions about services: that they are low-value-added, small-scale, and technologically unsophisticated industries. The chapter offers evidence that services such as communications, finance, transportation, and health care are large, capital-intensive industries responsible for commercial application of some of the most sophisticated technologies available. The main thrust of the chapter is an exploration of the ways in which technologies applied in services activities are changing the structure of domestic and global competition in both goods and services industries. Quinn's analysis of competitive structures—the way production and distribution are organized in different industries—makes a persuasive case that services and manufacturing activities are inextricably interdependent and that many of the opportunities in global manufacturing operations arise from technologies applied in services activities such as communications, transportation, and financial management. Those points are enriched and extended in the chapters by Ronald E. Kutscher and Faye Duchin.

The chapter by Kutscher on employment trends in services industries in the United States focuses considerable attention on explanations for the rapid growth of employment in producer services, services sold mostly to business rather than to individual consumers. Using a framework from input-output analysis, Kutscher estimates the growth in producer-services' output that would have taken place between 1972 and 1985 had there been no change in the structure of production. He finds that more than half of the growth in output in producer services between 1972 and 1985 can be attributed to changing business practices (changes in the structure of production). Pursuing explanations for this change in business practice, Kutscher explores the "unbundling" hypothesis—that producer services have grown as a result of manufacturers shifting in-house operations to external services providers. Based on analysis of employment data in manufacturing and in producer services, Kutscher concludes that "unbundling has been a very small factor in the employment growth of producer services." He offers alternative explanations for the rapid growth in producer services and closes his chapter with a discussion of the sustainability of a services-based economy.

The chapter by Duchin approaches the role of services in the U.S. economy from a fundamental level. Using input-output analysis, Duchin explores the requirements for manufactured goods in the production of services, the importance of services inputs for manufacturing production, the captive production of services by businesses, and the household production and purchases of services. The analysis, like the chapters by Quinn and Kutscher, illustrates the diversity of services activities and the degree of interdependence between services and manufacturing. Duchin concludes that

The growth of the services sectors has been accompanied by significant demand for construction, paper, transportation equipment, and various special-purpose capital goods that are among the largest inputs (in value) to virtually all the services sectors. The notion that services involve essentially people (and computers) turns out to be an unrealistic basis for policy.

The chapters by Quinn, Kutscher, and Duchin illustrate that an important economic and technological synergy exists between services and manufacturing; the services surrounding manufacturing are as much at the heart of manufacturing productivity as is the introduction of new, more efficient production machines. The information management surrounding manufacturing operations—just-in-time inventory techniques, new approaches to quality control, and systems for product delivery to market—represents only some of the services dimensions of manufacturing (see Quinn, 1988, for further discussion). Recognition of the interdependence of manufacturing and services, when combined with an understanding of the technological opportunities inherent in services activities, leads naturally to an interest in the character and level of productivity growth in services industries. In a very direct way, the performance of services industries affects the performance of the entire economy.

PRODUCTIVITY GROWTH IN SERVICES

Productivity growth statistics attempt to measure the performance of an economy or industrial sector in increasing output per unit of resources used. As John Kendrick and Jerome Mark explain in their chapters, productivity has grown slowly in services industries such as hotels and motels (an average of 0.4 percent per year between 1973 and 1985) and intercity trucking (0.4 percent per year between 1973 and 1985). In other services industries the rate of growth in productivity has been rapid. Productivity growth in telephone communications services averaged 6.2 percent per year between 1973 and 1985; in air transportation the average increase was 3.9 percent, and in gasoline service stations the average annual increase was 3.2 percent. In contrast the average annual rate of increase for all manufacturing industries over the period 1948–1985 was 2.7 percent.

In addition to reviewing different methods of measuring productivity growth Kendrick addresses policy options to increase productivity in services—policies that include supporting research and development for services, improving worker skills, or encouraging investment to bring the newest generation of technology into the workplace. Chapters in the companion volume

(Guile and Quinn, 1988), without directly addressing the productivity of industrial sectors, illustrate the role that technological advance can play in the productivity growth of sectors such as communications, automotive repair, and distribution (Davis, 1988; Fellowes and Frey, 1988; Larson, 1988).

Stephen S. Roach, in his chapter on investment and productivity growth in services, examines the problems of low productivity growth in services industries. According to his analysis the investment in information technologies by many services industries has been substantial over the past 20 years. Since 1965 services-producing industries have consistently been responsible for more than 50 percent of national capital spending, and an increasingly large share of the capital stock of services industries is in information technology. The puzzle remains, however, as to why this substantial investment in information technology has not produced more rapid productivity growth in services. Roach offers several possible explanations, including poor management of technology by services businesses.

One issue that becomes clear from the chapters by Roach and Kendrick is that the ability to grapple with policy questions regarding productivity in the services industry is badly hampered by serious definitional and measurement problems. The chapter by Mark, therefore, addresses a central issue for development of policies about services industries: data collection and measurement.

In manufacturing industries, productivity has a long history of measurement based on labor hours, capital expenditures, materials costs, and outputs. Accurate measurement of the inputs and outputs in manufacturing is difficult, but it pales in comparison with the measurement problems for services industries. For example, how can one measure the productivity of the real estate equity investment department of a financial institution? Is the output of the department represented by the number of investments, the dollar value invested, or the return on investment (something that is not usually known until several years after the investment is made)? Are the typical measured inputs—space, equipment, and labor hours consumed by the department in doing its business—adequate descriptors of the "production" of quality investments? Measures that deal with physical inputs and outputs may show the impact of changing technology (office automation, for example), but they do not relate well to meaningful definitions of productivity in the services industry of real estate investments.

In short, the character of production of many services makes the measurement of inputs and outputs problematic. Additionally, technological changes in products and process, changes in the organization of production, and national data-collection efforts that have historically been weak in the services area also act to confound meaningful measurement of productivity growth in services (see Helfand et al., 1984, and National Research Council, 1986). What is encouraging about the chapter by Mark is the degree to which the

Bureau of Labor Statistics has demonstrated that, with sufficient resources and attention to detail, it is possible to measure productivity growth reliably even in some of the most elusive services industries.

Much work still needs to be done in developing better methods for understanding productivity growth and structural change in the economy. The relatively simple and important concept of sector-specific productivity growth often flounders on difficult-to-measure inputs and outputs and on changes in technology or business practices. As the chapters by Mark, Kendrick, and Roach illustrate, existing measures of productivity growth in services provide a less than complete understanding of a number of important questions. Why has the return on investment in information technology by services businesses (as measured by productivity growth statistics) been so low, and, if it is really low, why have firms continued to invest? How does one understand the performance of U.S. producers in relation to that of foreign producers in improving the productivity of software production? Does the wide variation in productivity growth rates in services industries reflect real differences in rates of efficiency improvement or is it an outcome of some aspect of measurement or interpretation? A particular problem from a microeconomic or trade policy perspective is that sector-specific measures of productivity performance do not reveal much about the character of interdependence between industries or about the way technological change or trade affects the structure of production.

SERVICES TRADE

Services industries are becoming globalized in the same manner as manufacturing industries and for the same reasons: labor cost differentials, market access considerations, capital mobility, and growing dispersion of intellectual resources. Indeed, technological advance has made some services industries—finance, for example—a paradigm of global industries. In addition to being global industries in their own right, some services industries play central roles in the globalization of the production and distribution of goods; services industries operate the most sophisticated and complex real-time communications systems, and they provide the financial lifeblood of the entire global economy. The long-term vision of industrial evolution provided in the chapter by Frederick W. Smith is an example of services both in international trade and as a facilitator of economic globalization. The concept of a just-in-time global economy—an economy in which communications and transportation technology allow the reduction of manufacturing, retail, and wholesale inventories—is a remarkably simple yet reasonable extension of trends that have been with us since the invention of the telegraph and the rise of the railroad.

Services industries depend for their vitality on a broad spectrum of intellectual resources. The low end of the spectrum is unskilled labor (e.g., the routine data entry operator), but the other end of services employment can depend on significant intellectual achievement. Some services industries, such as software development for computers, engineering, systems design and management, and research and development, draw on the highest level of intellectual resources. The development of effective global communications, combined with the growth abroad of intellectual capabilities essential to services industries, has facilitated the emergence of services providers in Asia and Europe that rival those in the United States in their ability to design, manage, and operate complex systems. For example, much engineering design for U.S. engineering construction firms is now done abroad, in Europe, Korea, Taiwan, and elsewhere. Software is another example. Increasingly, it is becoming a target for other nations interested in building a high-technology export industry. The People's Republic of China and India are good examples, and they are skilled in this area. Not only is their relevant labor less expensive, it is of high quality. It is noteworthy that, at the other end of the intellectual spectrum, the sources of some routine services tasks such as data entry are also being established across national boundaries.

Even if one takes into account the fact that current services trade has most likely been underestimated in official figures, services trade is still small relative to merchandise trade: in 1984 when U.S. merchandise exports and imports were $218 billion and $341 billion, respectively, the very highest estimates of U.S. services exports and imports were $91 billion and $74 billion, respectively (Office of Technology Assessment, 1986, p. 39). Nonetheless, services trade is growing in the world economy and often represents economic activity in the most technologically advanced and the highest wage sectors. Growing U.S. concern over trade, combined with the immediacy of the next round of General Agreement on Tariffs and Trade (GATT) negotiations, has generated a wealth of study and insight into trade in services (American Enterprise Institute and the Coalition of Services Industries, 1987; Congressional Budget Office, 1987; Office of Technology Assessment, 1986; Stalson, 1985; U.S. Trade Representative, 1984). The chapter by Rauf Gönenç addresses changes in the structure of both national and international services industries and the implications of those changes for services trade.

Unlike manufactured goods, the heterogeneity of services creates problems even at the level of simply understanding what constitutes services trade; tourism by foreigners in the United States, travel or shipment on U.S. carriers by foreign travelers and shippers, distribution of U.S. films and recorded music overseas, and international management consulting are all components of services trade. As the chapter by Gönenç shows, the most striking thing about services trade is the wide range of political and economic issues that affect services trade. A few examples will suffice to make the point.

Services such as finance, transportation, and communications are heavily regulated in most countries. As a result, despite international trends toward deregulation of services (as discussed by Gönenç), trade in services is often a clash between competing national regulatory systems. The discussion by Frederick W. Smith of international air transport regulation is a case in point. In contrast to manufacturing, the primary focus of trade negotiations for such services is the elimination of nontariff barriers.

International trade in services such as motion pictures and software requires effective international regimes for dealing with copyrights, trademarks, and patents. The issues of intellectual property protection are foremost and central to the trade policy agendas of most industrialized nations. Both individual services industries and national interests would most likely be well served by stronger international regimes to prevent expropriation of such intellectual property (U.S. International Trade Commission, 1988).

Trade in some services can be greatly affected by laws inhibiting the movement of people across national boundaries. Many "top-of-the-line" services are affected. Engineering, law, and consulting often confront laws restricting the establishment of foreign offices or the licensing requirements of professional practice.

Growing trade in services also raises complex issues of national competitiveness in a global marketplace. On the one hand, U.S. consumers presumably benefit from fierce competition among services producers, whether those producers are based in the United States or in other nations. Also, provision of services by foreign-based producers may not have the same negative employment impacts on the U.S. economy as importing manufactured products. Transportation from Boston to Los Angeles can be provided only in the United States, whatever the ownership of the airline, and the demand for fast food can also be satisfied only locally.

On the other hand, increasing foreign direct investment in the United States in services (ownership of domestic businesses by individuals and companies from other nations; see data and discussion in the chapter by Quinn) raises complex questions about the U.S. national interest in industrial evolution. As discussed in the chapter by Richard W. Wright and Gunter A. Pauli, the growing presence of Japanese financial services institutions in U.S. and European markets brings forth national concerns about sovereign control of national financial markets and the sustainability of domestically based financial institutions. The same set of concerns would apply, presumably, were the U.S. air transport market to become dominated by foreign carriers. Few large industrialized nations are likely to embrace a dominant foreign-based presence in domestic transportation, communications, or financial markets. The national security concerns are obvious, but the economic security questions are just starting to emerge for the United States.

There is widespread agreement that some basic services strongly affect overall comparative economic performance. Electric utilities, health care, education, communications, transportation, and financial services are central elements of national economic development and welfare. Most important, however, from the point of view of international competition, they are the enabling industries—industries that, if efficiently provided, allow the functioning of an industrialized economy. They are prime targets for efficiency improvements in industrialized nations.

In the United States, the road, airport, and waste disposal systems are inadequate to meet current and projected demand (Herman and Ausubel, 1988) and are obvious targets for large-scale infrastructure development and deployment programs that would be of significant value in long-term U.S. economic performance. The difficult policy questions revolve around how a development is best paid for and which investment is most important now. Can some large services infrastructure investments, such as communications systems, airports, air traffic control systems, or waste treatment, be handled adequately by the private sector? If services infrastructures are provided by private investment, a new question arises: Is it desirable (or even possible) to distinguish between domestic and foreign capital investments in services infrastructures? Additionally, there are a number of unanswered questions about the strategic, long-term commercial importance of a strong technological or marketplace position in certain services industries.

In a global manufacturing business, success may come as much from the development and operation of a component ordering and control system as from efficient assembly operations. Although the logic is clear from the perspective of a corporation, the argument has yet to be worked out or agreed on from a national perspective; there is no obvious path between open competition among services in the international market and the desire for at least loose national control of certain core services. In short, aggressive international market competition in services industries such as finance, transportation, and communications is an important new reality in the global marketplace, one that many nations may be ill prepared to accept or adapt to.

KEY POLICY ISSUES AND DIRECTIONS
FOR FUTURE POLICY RESEARCH

The chapter by James Brian Quinn and Thomas L. Doorley addresses the central policy initiatives needed to support the effective use of technology in services industries. In particular, the chapter focuses on (1) macroeconomic and tax policies, (2) policies for investment (public, private, or mixed investment) in services infrastructures, (3) changes in the form of economic regulation, (4) human resources development policies, and (5) recognition of the interdependence of services and manufacturing in trade negotiations. The chapter integrates issues of productivity, investment, regulation, and trade in services to argue for more balanced attention to the role and importance of services industries in policy agendas.

Although individual chapters suggest a number of areas for further policy research, at least two important policy questions arise from consideration of the volume as a whole. First, the volume illustrates the depth and complexity of questions about the role of different services sectors, the effect of technological change on services, the interaction of services and manufacturing, and the impact of traded services on a national economy. The policy debates concerning U.S. competitiveness rarely reflect much understanding of the role and importance of services. This problem derives not only from the complex character of the issues but also from the weaknesses of available measures of U.S. economic structural change. One of the primary long-term policy challenges, therefore, is to pursue a more effective and agreed on analytical language for discussing technological and economic structural change and the implications of those changes for national policy.

Second, the chapters in this volume argue in favor of a serious and wide-ranging reconsideration of the role of the federal government in fostering research, development, and engineering related to services industries. It is clear that U.S. services industries need to remain abreast of, and be prepared to introduce, new technology from whatever source it may be available. Services firms have to increase their capabilities to conduct and—equally important—to exploit research and development. In recent years, concerns over the technological base of U.S. manufacturing industries have stimulated new forms of organization of research and development in manufacturing industries. Some of these mechanisms have been established in the private sector as a means of bringing to bear a critical mass of intellectual and financial capital in commercially strategic areas. The Microelectronics and Computer Technology Corporation in Austin, Texas, is a good example. Other important actions have involved the federal government. The Engineering Research Centers, established by the National Science Foundation, and Sematech (the semiconductor industry's partnership with the federal government) are examples of the trend.

In some services industries, joint efforts to support research and development are well established; the Electric Power Research Institute and the Gas Research Institute are good examples.Whether the technological base of a services industry can best be organized within single companies or as cooperative ventures will vary with the particular industry and the character of technological opportunities. It is worth further study, however, as to whether the federal government could play a more catalytic or supportive role in strengthening the national research and development base focused on services industries.

Finally, the political apparatus has not been excited much about the outlook for services, and that constitutes the primary policy challenge presented by this volume: to inform those responsible for national policy of the crucial issues facing the services industries and of the central role of services industries in maintaining the competitiveness of the U.S. economy.

ACKNOWLEDGMENTS

I would like to thank Robert M. White, president of the National Academy of Engineering, for his early insights about the policy questions raised by growing services industries and Harvey Brooks and Jesse H. Ausubel for their helpful comments on earlier drafts of this introduction to the volume.

REFERENCES

American Enterprise Institute and the Coalition of Service Industries. 1987. Trade in Services, Open Markets and the Uraguay Round Negotiations. Proceedings of Conference on Trade in Services and the Uraguay Round, Washington, D.C., November 18.

Bluestone, B., and B. Harrison. 1982. The Deindustrialization of America. Plant Closings, Community Abandonment, and the Dismantling of Basic Industry. New York: Basic Books.

Bluestone, B., and B. Harrison, 1986. The Great American Job Machine: The Proliferation of Low Wage Employment in the U.S. Economy. A study prepared for the Joint Economic Committee, December.

Chandler, A. D., Jr. 1977. The Visible Hand: The Managerial Revolution in American Business. Cambridge, Mass.: The Belknap Press of Harvard University Press.

Cohen, S. S., and J. Zysman. 1987. Manufacturing Matters: The Myth of the Post-Industrial Economy. New York: Basic Books.

Congressional Budget Office. 1987. The GATT Negotiations and U.S. Trade Policy. Report prepared for the Subcommittee on International Trade of the U.S. Senate Finance Committee. Washington, D.C.: U.S. Congress.

Cremeans, J. E. 1985. Three measures of structural change. Pp. 47–72 in The Service Economy: Opportunity, Threat or Myth? Proceedings of Workshop on Structural Change sponsored by the U.S. Department of Commerce Under Secretary for Economic Affairs, Washington, D.C., October 22.

Cyert, R. M., and D. C. Mowery, eds. 1987. Technology and Employment: Innovation and Growth in the U.S. Economy. Committee on Science, Engineering, and Public Policy. Washington, D.C.: National Academy Press.

Davis, J. H. 1988. Cellular mobile telephone service. In Managing Innovation: Cases from the Services Industry, B. R. Guile and J. B. Quinn, eds. Washington, D.C.: National Academy Press.

Fellowes, F. A., and D. N. Frey. 1988. Pictures and parts: Delivering an automated automotive parts catalog. In Managing Innovation: Cases from the Services Industry, B. R. Guile and J. B. Quinn, eds. Washington, D.C.: National Academy Press.

Guile, Bruce R., and James Brian Quinn, eds. 1988. Managing Innovation: Cases from the Services Industries. Washington, D.C.: National Academy Press.

Helfand, S. D., V. Natrella, and A. E. Pisarski, eds. 1984. Statistics for Transportation, Communication, and Finance and Insurance: Data Availability and Needs. Staff paper prepared for the Committee on National Statistics, National Research Council. Washington, D.C.: National Academy Press.

Herman, R., and J. H. Ausubel. 1988. Cities and infrastructure: Synthesis and perspectives. Pp. 1–21 in Cities and Their Vital Systems: Infrastructure Past, Present, and Future, J. H.

Ausubel and R. Herman, eds. National Academy of Engineering. Washington, D.C.: National Academy Press.

Larson, R. C. 1988. Operations research and the services industries. In Managing Innovation: Cases from the Services Industry, B. R. Guile and J. B. Quinn, eds. Washington, D.C.: National Academy Press.

Lawrence, R. Z. 1984. Can America Compete? Washington, D.C.: The Brookings Institution.

Mishel, L. 1988. Manufacturing Numbers: How Inaccurate Statistics Conceal U.S. Industrial Decline. Washington, D.C.: Economic Policy Institute.

Moody, G. 1985. Growth of the service industries: A description and assessment. Pp. 73–104 in The Service Economy: Opportunity, Threat or Myth? Proceedings of Workshop on Structural Change sponsored by the U.S. Department of Commerce Under Secretary for Economic Affairs, Washington, D.C., October 22.

National Research Council. 1986. Statistics About Service Industries: Report of a Conference. The Committee on National Statistics, Commission on Behavioral and Social Sciences and Education. Washington, D.C.: National Academy Press.

Office of Technology Assessment. 1986. Trade in Services. Exports and Foreign Revenues. Report prepared for the U.S. Senate Committees on Governmental Affairs and Foreign Relations and the U.S. House of Representatives Committee on Small Business. Special Report NO. OTA-ITE-316. Washington, D.C.: U.S.Congress.

Office of Technology Assessment. 1988. Paying the Bill: Manufacturing and America's Trade Deficit. Washington, D.C.: U.S. Congress.

Quinn, J. B. 1988. Services technology and manufacturing: Cornerstones of the U.S. economy. In Managing Innovation: Cases from the Services Industry, B. R. Guile and J. B. Quinn, eds. Washington, D.C.: National Academy Press.

Stalson, H. 1985. U.S. Service Exports and Foreign Barriers: An Agenda for Negotiations. Report No. 219. Washington, D.C.: National Planning Association.

U.S. International Trade Commission. 1988. Foreign Protection of Intellectual Property Rights and the Effect on U.S. Industry and Trade. USITC Publication 2065. Washington, D.C.: U.S. Government Printing Office.

U.S. Trade Representative. 1984. U.S. National Study on Trade in Services. A Submission by the United States Government to the General Agreement on Tariffs and Trade. Washington, D.C.: U.S. Government Printing Office.

Technology in Services:
Past Myths and Future Challenges

JAMES BRIAN QUINN

In recent years many have expressed concerns over the decline of U.S. manufacturing. Yet most government statistics show that total employment in manufacturing has decreased only marginally from long-term levels, and both real gross national product (GNP) and real net value added attributable to manufacturing grew steadily until the mid-1980s as shown in Figures 1–3. Although some recent publications have challenged the assumptions behind these statistics (Mishel, 1988; Office of Technology Assessment, 1988), most projections through the year 2000 show manufacturing output growing at a rate similar to that of the total economy (Personick, 1987). Nevertheless, individual manufacturing industries—such as automobiles, basic steel, and consumer electronics—have suffered real declines in employment and output over the last 15 years, and in some fields U.S. goods have become uncompetitive in world markets.

Meanwhile, the services sector has grown steadily and now accounts for 71 percent of the country's GNP and 75 percent of its employment (see Table 1). The movement toward services has been a long-term trend not only in the United States but also in all major industrialized countries. It is time to put aside some popular misconceptions about services that may have fit better in an earlier period, but can now seriously impair our capacity to meet future challenges. Those who fail to understand the realities and potentials of services are very likely to mismanage their own enterprises. They will certainly support some poor national policy choices.

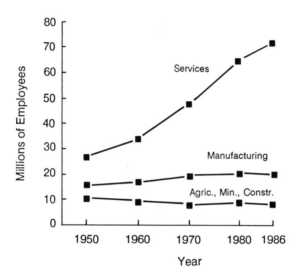

FIGURE 1 Employment trends by sector. SOURCE: Bureau of Economic Analysis, The National Income and Product Accounts of the United States; Bureau of Labor Statistics, Establishment Data Base: Employees on Nonagricultural Payrolls by Major Industry.

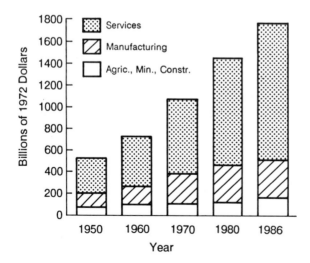

FIGURE 2 Real gross national product by sector. SOURCE: Bureau of Economic Analysis, The National Income and Product Accounts of the United States.

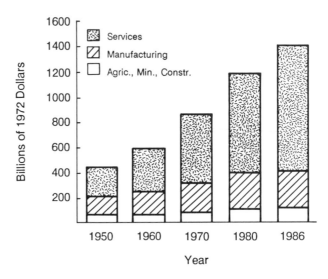

FIGURE 3   Real value added by sector. SOURCE: Bureau of Economic Analysis, The National Income and Product Accounts of the United States.

DISPELLING PAST MISCONCEPTIONS ABOUT SERVICES

Unfortunately, many executives and policymakers have tended to dismiss services as predominantly "taking in laundry" or "making hamburgers" for each other. Such simplifications belie the complexity, technological sophistication, and continuing growth potentials of services in the 1980s and 1990s. While there is not a complete consensus on definitions, most authorities consider the services sector to include all economic activities whose output (1) is not a product or construction, (2) is generally consumed at the time it is produced, and (3) provides added value in forms (such as convenience, amusement, timeliness, comfort, or health) that are essentially intangible concerns of its purchaser (Kutscher and Mark, 1983). *The Economist* has more simply defined services as "anything sold in trade that could not be dropped on your foot."

With these definitions in mind, one can readily reassess some of the most misleading myths, held over from the past, about services.

The "lower value" misconception—perhaps first stated by Adam Smith— regards services as somehow less important on a "human needs scale" than products. Because services are essentially marginal (so the argument goes), they cannot add the same economic value or provide the growth potentials that manufactures can.

Perhaps in elemental societies it is true that the first production of food,

TABLE 1   U.S. Gross National Product and Employment by Industry, 1986

	GNP ($ billion)		Employment (millions)	
TOTAL ECONOMY	$4,235		108.0	
Goods				
Agriculture, forestry, fisheries	93		1.7	
Mining and construction	293		5.7	
Manufacturing	825		19.1	
Total Goods	$1,211	[29%]	26.5	[25%]
Private Services				
Finance, insurance, real estate	695		6.5	
Retail trade	408		18.4	
Wholesale trade	295		5.8	
Transportation and public utilities	276		4.0	
Communications	115		1.3	
Other services	700		24.9	
Total Private Services	$2,489	[59%]	60.8	[56%]
Government and government enterprises	507		20.6	
Total Services	$2,996	[71%]	81.4	[75%]
Rest of world and statistical discrepancy	30			

NOTE: The services sector includes some giant industries, larger than the great manufacturing industries such as automobiles or steel. Even the "other private services" category contains such sizable activities as health services ($199 billion), education ($27.0 billion), entertainment ($30.0 billion), design engineering, consulting, legal ($52.0 billion), software, and architecture professions that are sophisticated generators and users of technology. Large government services groups, such as the Department of Defense, the National Aeronautics and Space Administration, the Department of Agriculture, or the Department of Energy, support other huge, advanced technology businesses.
SOURCE: Bureau of Economic Analysis, The National Income and Product Accounts of the United States (July 1987, Table 6.1, p. 57; Table 6.6B, p. 60).

basic shelter, or clothing may take precedence over other demands. However, as soon as there is even a local self-sufficiency or surplus in a single product, the extra production has little value without further distribution, financing, or storage—all "services" activities. In most emerging societies, services such as health care, education, trading, entertainment, religion, banking, law, and the arts quickly become more highly valued (high priced or capable of generating great wealth) than basic production. Such differentials tend to be even more marked in affluent societies.

Far from being inferior economic outputs, services are directly interchangeable with manufactures in a wide variety of situations. Few customers care whether a refrigerator manufacturer implements a particular feature through a hardware circuit or by internal software. New computer-aided design and manufacturing software can substitute for added production or design equipment, and improved transportation or handling services can lower a manufacturer's costs as effectively as cutting direct labor or materials inputs. These "services" can improve productivity or add value just like any new investment in physical handling machinery or product features.

Even more fundamentally, products themselves are only physical embodiments of the services they deliver. A diskette delivers a software program or data set. An automobile delivers flexible transportation—a service. Electrical appliances deliver entertainment, dishwashing, clothes cleaning or drying, and convenient cooking or food storage—all services. In fact, most products merely provide a more convenient or less costly form in which to purchase services. It should be no surprise then to find that, nationally, the total value added in services industries is higher than that in manufacturing (see Figure 3). Although total value added per employee in services is lower nationally than that in manufacturing, in the strategic business units of the larger companies sampled by Profit Impact of Market Strategy (PIMS) data, value added per services employee is very comparable to that per manufacturing employee (see Figure 4). This suggests that there are significant economies of scale in the larger services enterprises.

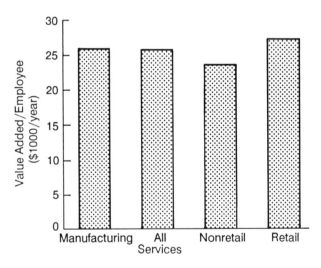

FIGURE 4   Profit Impact of Market Strategy (PIMS) indices of value added. SOURCE: Figures 4–6 were prepared by Christopher Gagnon from the PIMS 1985 data base of the Strategic Planning Institute, Cambridge, Mass. PIMS data are provided by strategic business units from cooperating companies on a voluntary basis. Most of the cooperating companies are major U.S. corporations.

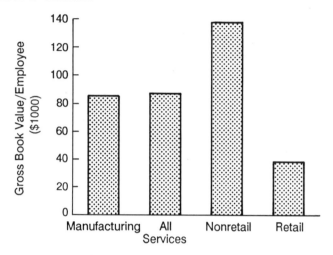

FIGURE 5 Profit Impact of Market Strategy (PIMS) indices of capital intensity. SOURCE: See Figure 4 legend.

The "low capital intensity" perception asserts that services industries are less capital intensive and much less technologically based than manufacturing. While this may be so for small scale retailing and domestic services, many services sectors today are very capital and technologically intensive. The prime examples are communications, transportation, pipelines, and electric utilities. However, the banking, entertainment, health care, financial services, auto rental, message delivery, and retailing industries also increasingly qualify.

In his chapter, Roach calculates that total capital investment—and in particular high-technology investment—per "information worker" (services activities) has been rising rapidly since the mid-1960s and now exceeds that for workers in basic industrial activities (Roach, 1985, 1988). Similarly, Kutscher and Mark (1983) show that nearly half of the top 30 most capital-intensive industries (of 145 studied) were services. Furthermore, certain services industries—notably railroads, pipelines, broadcasting, communications, public utilities, and air transport—were among the most capital intensive of all industries. Surprisingly, few services industries were found in the three lowest capital intensity deciles. Our analyses, based on PIMS data, also show aggregate capital intensity in larger services enterprises comparable to those in manufacturing (see Figure 5).

The "small scale" misconception considers the services sector as too small in scale and too diffuse either to buy major technological systems or to do research on its own. Although complete Hirfendahl indices of concentration are not available, PIMS and Fortune 500 data suggest that concentration and scale among larger services units are comparable to those in larger manufacturing units (see Figure 6). Many companies in the services sector clearly have the financial power to buy technology as it becomes available and needed.

Large banks, airlines, utilities, financial services institutions, communications companies, and hospital, hotel, or retail chains not only have the scale needed to purchase technology, but can also contribute to its long-term creation, reduction to practice, and introduction. These and other companies acting in intermediary roles (like distributors) often then push technology out into smaller services companies, assisting or forcing its rapid diffusion. Many of these large institutions now support major research programs, creating and guiding new technological developments themselves, as Citicorp did with automated teller machines (ATMs) and Federal Express has done with its COSMOS, DADS, and mass sorting systems.

The "services can't produce wealth" viewpoint holds that services are not capable of producing the ever higher levels of real income and personal wealth that have been the hallmarks of the "industrial" era. This argument assumes, in part, that services inherently cannot achieve the productivity increases available through automation in manufacturing. If not, services cannot possibly achieve the inflation-free income growth rates manufacturing can.

Productivity in services is notoriously difficult to measure because of the

FIGURE 6   Profit Impact Market Strategy (PIMS) indices of concentration. SOURCE: See Figure 4 legend.

many problems, specified in the chapter by Mark, in numerically defining output units and quality differences in services (Mark, 1986, 1988). For example, how does one evaluate medical procedures that may use fewer resources, but substantially increase patients' pain or morbidity levels? Is the number of letters delivered per mail worker a meaningful productivity measure if more letters are late or lost? Of what use are standard economic productivity measures that ignore the output value of public services such as sewage treatment plants or assume that this output value is only equal to its cost? One should be very cautious in interpreting aggregate productivity data about services.

Productivity measures are more valid in competitive arenas where customers can make direct purchase trade-offs between one class of services and other services or manufactures. Here the sales value of a service can provide a better surrogate for measuring output. In some of these more measurable segments, Bureau of Labor Statistics (BLS) historical data suggest that individual services industries can sustain productivity growth rates as high as those in manufacturing for substantial periods (see Table 2).

Although the 1980–1985 aggregates are less encouraging, those for 1985–1986 have rebounded, and data for more detailed Standard Industrial Classification (SIC) categories show wide variations in productivity growth rates among different segments of both manufacturing and services (see Table 3). Both aggregate and narrower industry-oriented statistics suggest that there is no inherent reason why many services industries cannot keep up with—or outperform—individual product industries in terms of productivity increases. One key will be the ability and willingness to make new technology investments in each sector.

Manufacturing productivity has improved markedly since 1980, based on extensive restructurings and investments. However, new technology investments by services producers have also soared since 1975, with financial services, wholesale trade, and health services among the leaders (Roach, 1987). Interestingly, the information technologies at the heart of services investments often offer multiplying benefits. As workers master the new technologies, they frequently discover or create new or more productive applications not envisioned in the original investment. Both output values and their own skills—human capital values not included in productivity measures—are enhanced. After a delay for absorption and employment adjustments, many experts anticipate that these expenditures will boost services productivity substantially, as suggested by Kendrick in his chapter, although the results may not show up as dramatically in aggregates because of the measurement problems cited above.

In a similar vein, Barras (1986) in a recent and very thorough study of the British economy showed the services sector productivity growing at 2.9 percent annually (based on the real value of output per employee) from 1960

TABLE 2 Productivity Changes by Industry Group

	Compound Annual Rates of Change (Output per hour, all persons)				
	1950–1979	1950–1969	1970–1979	1980–1985	1985–1986
GOODS-PRODUCING INDUSTRIES (TOTAL)	2.5%	3.2%	1.2%	2.7%	2.1%
Manufacturing	2.4	2.4	2.5	3.7	3.7
Durable	1.9	2.0	2.2	5.3	5.2
Nondurable	3.0	3.0	3.1	2.2	1.4
SERVICES-PRODUCING INDUSTRIES (TOTAL)	2.1	2.4	1.5	0.7	2.0
Transportation	2.1	2.3	2.0	-1.0	3.6
Communications	4.9	5.0	4.7	3.8	3.3
Utilities	4.5	6.0	2.1	2.2	-3.0
Wholesale trade	2.4	2.9	1.4	2.7	5.0
Retail trade	1.8	1.9	1.7	1.7	3.6
Finance, insurance, and real estate	1.1	1.5	0.5	-0.9	1.5
Other services	1.9	2.3	0.9	-0.1	0.6
Government enterprises	0.2	0.0	0.6	-0.7	0.8

SOURCE: Office of Productivity and Technology, Bureau of Labor Statistics (1987).

TABLE 3   Productivity Changes by SIC Code

Compound Annual Rate of Change in Indices of Output per Employee Hour,[a] 1980–1985 (%)	Manufacturing Industries		Services Industries	
	Number	(%)	Number	(%)
Greater than 6.00	7	8.0	3	13.6
5.00 to 5.99	6	6.8	1	4.5
4.00 to 4.99	9	10.2	2	9.1
3.00 to 3.99	12	13.6	4	18.2
2.00 to 2.99	14	15.9	2	9.1
1.00 to 1.99	15	17.0	1	4.5
0.00 to 0.99	7	8.0	2	9.1
−0.01 to −0.99	7	8.0	3	13.6
−1.00 to −1.99	7	8.0	3	13.6
−2.00 to −2.99	2	2.3	1	4.5
−3.00 or less	2	2.3	0	0.0
Totals	88	100.1	22	99.8

[a]1977 = 100.
SOURCE: Bureau of Labor Statistics (October 1, 1987). (Percentages do not total 100 due to rounding.)

to 1981, whereas manufacturing productivity grew at less than 1 percent on the same basis. The primary causes of this phenomenon were (1) the continued demand growth in services, which led to (2) both capital deepening (in the sheer quantity or accumulation of capital) in services and improved capital quality (output gain per unit of invested capital) within services. The analyses of Barras (1986), Kendrick (1987), and Kutscher and Mark (1983) all suggest that shifts to services have not been the most important culprit in slowdowns of U.S. or British national productivity growth.

What about the longer term prospects for growth based on services? Because the value of all products or services is created solely in the mind (i.e., a jewel, an opera, a Ferrari, a sightseeing tour, or a stylish coat may have little functional value relative to its high price), the growth of a services economy is limited only by the capacity of the human mind to conceive of activities as having high utility. Surely, a safer, healthier, better educated, more stable society can easily be considered "wealthier" than one with more physical goods. Moreover, this wealth can be passed on to future generations. Services, such as better education, art, music, literature, information repositories, public health levels, mechanical skills, scientific or design knowhow, and institutions of law represent critical investments for the future, yielding higher productivity and living standards both in the present and in the future. In fact, such dimensions have generally been considered both the most important contributions to and measures of a nation's wealth throughout history.

HOW TECHNOLOGY AFFECTS COMPETITIVE STRUCTURES

Why have services gained such importance in recent years? Steady productivity increases in agriculture and manufacturing, largely induced by technology, have meant that it took ever fewer hours of work to produce or buy a pound of food, an automobile, a piece of furniture, or a home appliance. Whereas productivity improved, demand for goods was somewhat capped; people consume only so many pounds of food, automobiles, sofas, or washing machines. The relative utility of other possible purchases therefore went up for each individual. Also, in recent years, people have begun to want more services. Thus, to a large extent, the shift to a services economy has been a natural outgrowth of productivity increases induced by technological advances in the product sector. Growth in services sector employment has more than offset declines in the goods-producing sector (see Figure 1).

Simultaneously, new technologies also vastly improved performance in virtually all services sectors. Jet aircraft made long haul passenger and freight handling much more efficient and convenient. New noninvasive imaging devices, drugs, diagnostics, life support systems, and surgical procedures revolutionized medical practice. New containerization, loading, refrigeration, and handling techniques for volatile liquids, by making it possible to transport virtually all goods safely and effectively, vastly extended international trade. Electronics, information, and communications technologies stimulated new innovations in virtually all services areas—most notably in retailing and wholesale trade, engineering design, financial services, communications, and entertainment.

The restructurings caused by such technologies have been detailed elsewhere (Quinn, 1987). Major new technologies in services seem to generate certain distinctive and repetitive impact patterns:

• New economies of scale appear which cause many services activities to centralize into larger institutions, at first concentrating activities into fewer large units, then allowing renewed decentralization as smaller units in more dispersed locations link into networks with larger companies and deliver services to widely dispersed locations. This pattern has recurred in health care, air transport, insurance, ground transport, banking and financial services, and communications. Midsized services enterprises, unable to afford the new technologies themselves, have often been forced to merge upward in scale, or niche radically, or to go out of business. Strategists refer to this pressure on midsized firms as "being caught in the middle" (Porter, 1985).

• New economies of scope frequently provide powerful second-order effects. Once properly installed, the same technologies that created new scale economies allow services enterprises to handle a much wider set of data, output functions, or customers without significant cost increases and often with cost decreases through allocating technology development or equipment costs over a richer base of operations. Thus, banks (Citicorp), airlines (American), retailers (Sears), and travel-bank services (American Express) use their installed facilities and networks to extend their presence into a broad range of new activities. Such economies become so extensive that even very large companies like IBM have established partnering relationships with others such as NTT, MCI, Merrill Lynch, Rolm, British Telecom, and Mitsubishi to create worldwide electronics and equipment networks (*The Economist*, 1985b, p. 36).

• Increased complexity can often be handled economically by the new technologies. Electronic systems and computer models have been the main enabling technologies—but are by no means alone—in permitting management of much greater complexity. A variety of new sensing, telecommunications, information handling, materials, and processing technologies now routinely: design, build, and test radical new designs for boat hulls and aircraft; specify structures for new molecules and predict their performance; suggest and test hypotheses for medical research; access and analyze global and astronomical data bases; run remote factories and processes; handle worldwide monetary and securities transactions; control effluents and water supplies; monitor environmental and political events; and manage huge transportation systems with a precision and speed previously impossible.

One more complete example will dramatize some of the ways services technologies—and these three modes of impact—can interact to restructure competitive relations within and among modern nations:

In the mid-1960s and early 1970s securities trading houses found that they were being overwhelmed by the need to physically handle the 10–12 million shares then being transferred daily. The major Wall Street firms formed what became the Depository Trust Company to immobilize virtually all traded certificates under one roof and to automate what had been the back office handling of such certificates. This technology forced many small and midsized brokerage houses, which could not afford their own automation, to merge or affiliate with the large houses.

As the surviving houses developed their own in-house electronics systems for automated trading, they found they could not only handle increased transaction volumes (now sometimes over 500 million shares per day) but could also introduce a variety of new products such as cash management accounts tailored to customers' needs. To exploit these potentials, they developed networks of decentralized local offices all over the country. As European exchanges and brokerage houses later automated themselves, world financial markets became so integrated that countries began changing their trading rules to maintain national competitiveness. Large industrial companies began to place their own securities (commercial paper) directly on world markets, and the finances of all advanced economies have now become intimately interlinked in ways that have changed the very nature of international competition. More on this later.

SERVICES TECHNOLOGIES AND THE FUTURE

In recognizing how powerful these past patterns have been, what are some of the more probable and important consequences of services technologies in the future?

Changed Employment Patterns

Perhaps the most immediate concerns have to do with job potentials, employment patterns, and future growth capabilities in services. Despite the political attention given to manufacturing and the very real problems of certain industries, every state now has more services jobs than it has manufacturing, mining, construction, and agriculture jobs combined. No state suffered a decline in services jobs from 1976 to 1986. On the other hand, manufacturing jobs shifted markedly among both industries and regions: 12 states lost more than 10,000 such jobs, but 15 added more than 10,000 manufacturing jobs during this period. Jobs in the total goods-producing sector grew by 1.2 million over these years, almost all in construction. However, services provided the critical engine of employment growth for the country and all its geographical areas, generating 16.9 million jobs, or 93 percent of all private employment increases from 1976 to 1986.[1]

The real question is what kind of jobs are being created? Both historical data and the projected fastest growing occupations of the 1986–2000 era indicate a more professional and white-collar mix of jobs (see the chapter by Kutscher and Table 4). All of the top 25 most rapidly growing occupations

TABLE 4   Total Civilian Employment

Employment	Numbers (millions)		Percent Change	
	1986	2000 (estimated)	1972–1986	1986–2000
TOTAL	111.6	133.0	33.4	19.2
Executive, administrative, managerial	10.6	13.6	73.4	28.7
Professional workers	13.5	17.2	57.5	27.0
Technical and support	3.7	5.1	74.5	38.2
Salesworkers	12.6	16.3	54.6	29.6
Administrative support and clerical	19.9	22.1	35.2	11.4
Private household workers	1.0	1.0	− 31.9	− 2.7
Services workers, nondomestic	16.6	22.0	45.9	32.7
Precision, craft, repair	13.9	15.6	29.6	12.0
Operators, fabricators, laborers	16.3	16.7	− 1.3	2.6
Farming, forestry, fishing	3.6	3.4	− 10.4	− 4.6

SOURCE: Silvestri and Lukasiewicz (1987, Table 1, p. 47).

to 2000 are services related including paralegal (104 percent growth), medical assistants (90 percent), data processing equipment repairers (80 percent), computer systems analysts (76 percent), employment interviewers (71 percent), radiology technicians (65 percent), operations and systems researchers (54 percent), security guards (48 percent), and so on (Silvestri and Lukasiewicz, 1987).

Although many of these jobs offer high wages, their numbers are somewhat limited. Overall data show average hourly wages in services to be lower than the average in manufacturing. However, wages in some individual services industries are higher than those in many manufacturing industries. Also, wages in several services industries are growing faster than those in the manufacturing sector (see Table 5).

There are conflicting forecasts of future implications. Bluestone and Harrison (1986) suggest that the new jobs in the U.S. economy may be in lower wage categories (and thus cause a decline in middle-income earners). However, Neal Rosenthal of BLS notes that over the decade 1973–1982 (during which there was a continued shift to services jobs) there was only a slight decline in middle-income jobs, but a much larger decline in lower-income jobs (Rosenthal, 1985). McMahon and Tschetter (1987) show that from 1983 to 1985 employment grew in the middle and upper thirds of the wage distribution while dropping in the lower third. It is difficult to reconcile the

TABLE 5   Average Hourly Earnings in Private Nonagricultural Jobs (Production or Nonsupervisory Workers)

Industry	1982	June 1987	Compound Average Annual Growth Rate, 1982–1987
TOTAL PRIVATE SECTOR	$7.68	$8.94[a]	3.1%
Mining	10.77	12.44	2.9
Construction	11.63	12.61	1.6
Manufacturing total	8.50	9.87	3.0
Durable	9.06	10.42	2.8
Nondurable	7.73	9.11	3.3
Transportation and public utilities	10.32	11.91	2.9
Wholesale trade	8.09	9.57	3.4
Retail trade	5.48	6.08	2.1
Finance, insurance, and real estate	6.78	8.68	5.1
Services (other)	6.92	8.35	3.8

[a]Seasonally adjusted.
SOURCE: Bureau of Labor Statistics, Establishment Data Base: Average Hours and Earnings of Production or Nonsupervisory Workers on Private Nonagricultural Payrolls by Major Industry (November 1983; November 1987).

disparate conclusions the authors seem to derive from the purportedly same data base. The chapter by Kutscher does not attempt to do so. Rather, it considers some of the probable causes and effects of these employment shifts. Whichever trend one accepts, the truly important questions are (1) whether in the future technology can improve productivity and value added in services more rapidly than in manufacturing, thus permitting higher future relative wages, and (2) whether wages in new services jobs will be higher than the wages employees in distressed manufacturing industries would have to accept to keep their companies from going overseas to achieve competitive costs. In either event, services employment and growth would permit a higher U.S. standard of living than would subsidizing certain mature manufacturing industries to save jobs there.[2]

Thurow (1987) notes that the widespread entry of women and the less skilled into new job openings, mostly services, has created some of the downward bias in services wages. However, by creating entry level jobs and increasing the availability of more flexible part-time jobs, services seem to have stabilized or increased incomes for the multiple-earner families so prevalent today (Levy, 1987). In addition, the working conditions of services jobs tend to be less physically difficult and hazardous than many in manufacturing, and services locations can be more dispersed and convenient, corresponding to suburban living habits. Thus, the quality of services jobs may be higher on two important scales, perhaps partially offsetting the attractiveness of manufacturing's somewhat higher wages.

Economic and Job Stability

Yet what about economic and job stability? If services activities really were marginal relative to manufactures, one would expect people to give up their services first during recessions, and thus exacerbate employment and economic problems. In fact, just the opposite happens. BLS data show that in business cycles from 1948 to 1980, U.S. services employment advanced at an average annual rate of 2.1 percent during economic contractions and 4.8 percent during expansions. In contrast, employment in the goods-producing sector declined at an average annual rate of 8.3 percent in recessions and advanced at a rate of only 3.8 percent in expansions (Urquart, 1981). A study by Moore (1987) indicates that employment in services is considerably more stable than that in the goods sector, as measured by overall percentage gains or losses during recessions (see Figure 7). Moore's study shows that, during the four recessions from 1948 to 1961, employment in private services dropped an average of only 1 percent while that in goods-producing industries plunged by 7.2 percent. During the four recessions from 1969 to 1982, private services employment actually rose an average of 1.0 percent while the goods-producing industries dropped an average of 7.9

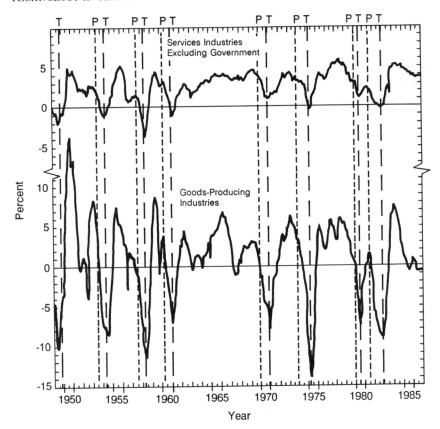

FIGURE 7   Growth rate in employment in the services industries and goods-producing industries. Growth rates are annual rates based on the ratio of the current month's employment to the average level of employment during the previous 12 months, by using seasonally adjusted data. Vertical lines are business cycle peaks (P) and troughs (T). SOURCE: Moore (1987, p. 16).

percent. Added services employment kept average total nonfarm employment losses in recessions to only 1.6 percent during this period. In Canada a similar effect occurred, with services employment declining only in the 1953–1954 and 1981–1982 recessions (Chand, 1986).

Why does this happen? A large portion of consumer spending on services is less discretionary than that for many products. During recessions, although people may go to the movies less often or purchase fewer personal services, they seem reluctant to give up their telephone, health care, education, banking, police, fire protection, and utility services. Instead they postpone durable goods expenditures (i.e., during recessions these goods are worth less at the margin than selected services). Stability is also enhanced because there is no pause in services production while inventories (nonexistent in most services) must be depleted. Further, income levels—rather than employment—tend to decline in services during recessions because a substantial portion of services incomes comes from part-time work, tips, commissions, or profit sharing. Finally, public transportation and services associated with population growth (such as personal security and health care expenditures) may actually increase during recessions.

Interestingly, therefore, the new "services economy" may well be more stable than an "industrial" economy. In addition to reducing the depth of recessions, services industry growth has helped make them shorter (Moore, 1987). Services industries are major purchasers of capital equipment, and the replacement of services capital items tends to be more constant because the useful life of equipment in services—dominated by the rapidly advancing information technologies—is often shorter. In addition, many services companies continue to grow and invest in capital goods during recessions and thus level out or pull up the manufacturing sector along with the total economy. Although the relative rate of investment in services during recessions, as compared with expansion periods, is an interesting question for further research, current data suggest that policymakers and executives seeking economic stability and new manufacturing markets should actively support—not disparage—the growth of services industries.

The Manufacturing-Services Interface

Many have noted that the services sector is very dependent on manufacturing, i.e., services units often provide transportation, finance, advertising, repair, distribution, or communications support for manufactured goods. Although a large number of these services would still be provided in the United States regardless of where the product is manufactured (e.g., selling a Toyota here requires many of the same U.S.-based services activities as selling a Ford), important design support functions and supplier interlinkage services could move overseas if manufacturing was not performed here.

On the other hand, a healthy manufacturing sector is probably equally dependent on services. Although the lagging data for U.S. input-output tables—and poor definition of sectors—do not allow complete current calculations, some specific studies suggest that 85 percent of the communications and related information technologies equipment sold in the United States (1985) went to information industries (dominantly services) (Roach, 1987), and 70 percent of all installed computer systems in Britain (1984) were in services industries (*The Economist*, 1985a). Furthermore, if 75 percent of all people are employed in services, clearly they and their enterprises must be the major markets for consumer and commercial products.

In addition, manufacturing competitiveness depends heavily on services efficiencies. Greater efficiencies in communications, transportation, financing, distribution, health care, waste handling, etc., markedly affect manufacturers' direct costs and the standard of living workers can achieve with their paychecks. In most areas, the efficiency of U.S. services industries has been more than competitive, even with Japan. Although statistics on output per hour are hard to evaluate and some changes have undoubtedly occurred, 1980 data indicate that except in the area of finance and insurance, the United States was significantly more efficient than Japan in major services sectors (see Figure 8). In fact, Japan has usually suffered a large negative trade balance in services (Tanaka, 1984).

However, two other aspects of the services-manufacturing interface may be even more crucial. First is the recognition that within manufacturing, 75–85 percent of all value added and a similar percentage of costs are due to services activities (Office of the U.S. Trade Representative, 1983; Vollmann, 1986, p. 149). The major value added to a product is typically due less to its basic commodity value (i.e., producing the "body in white" for an automobile or the grain and vegetables for a processed food) than to the styling features, perceived quality, etc., added by "services" activities inside

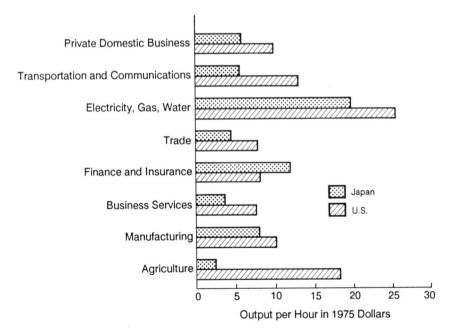

FIGURE 8 Productivity comparisons for 1980: United States versus Japan. SOURCE: UNIPUB (1984, p. 67).

or outside the producing company. In the wide range of manufacturing companies that we sampled—not just continuous process industries (such as oil refining or chemicals manufacturing)—planning, accounting, inventory, quality assurance, transportation, design, advertising, distribution, etc. costs generally outweigh direct labor costs by factors between 3 and 10 to 1. Yet, most companies have diligently driven out direct labor costs while only cautiously attacking the greater costs and value added potentials of services. Managing services activities for manufacturing enterprises and designing better measurement and control systems for them can provide a major attack point for improving competitiveness in the future.

Services technologies allow much more responsiveness to fluctuating or individualized demand patterns. Proper integration of these technologies into manufacturing and distribution could significantly increase the number and range of goods it pays to produce in the United States, rather than overseas. As personal affluence or the sheer size of markets grows, there is an increasing demand for differentiated, individualized, or customized products. Both flexible manufacturing and automated services technologies can help manufacturers exploit these trends. Manufacturing and wholesale-retail customers want both shorter or just-in-time (JIT) inventory responses and higher variability in products available. Consumers want delivery on products that reflect their particular tastes now, rather than in 3–12 weeks (Blackwell and Talarsyk, 1983; Seth, 1983).

As a result, successful manufacturing increasingly requires rapid feedback from the marketplace, more customized products, and accurate delivery in shorter cycle times. Some retailers, such as The Limited, daily aggregate sales data electronically from their entire retail network, transmit stock orders by telecommunications to manufacturer-suppliers all over the world, and require accurate deliveries within days (often by transpacific 747 flights) to their specific distribution points.

U.S.-based manufacturers able to link their market intelligence and flexible production systems directly to such customers' in-house networks should have very real competitive advantages (in timing and transportation costs) over foreign producers. In heavy, bulky, or complex products such as automobiles, it could even become impossible for overseas producers of more standardized lines to compete against this kind of responsive, integrated, manufacturing-retail system. Such considerations seem an important factor in Honda's rapid proliferation of automobile models and options, its increased emphasis on U.S. production, and the importation of its own JIT suppliers to the United States (when U.S. manufacturers proved unwilling or unable to meet Honda's services and quality demands). Honda now offers a much wider variety of styles and options and faster delivery to the U.S. marketplace than it ever could have from a solely Japanese base. Interestingly, it will have to source more of its high-value-added components and subassemblies from nearby U.S.-based producers to obtain the required response time without escalating inventory costs beyond economic levels. In addition, its productivity allows Honda to ship U.S.-made cars to Japan with exchange rates at current levels.

Changing International Competitive Structures

Increasingly, success in manufacturing will go to those who successfully combine the new potentials of services and manufacturing technologies. Manufacturer-distributors such as Sears or other chain discount stores are integrated from international supply sources to after-sale financial support. They can offer instant distribution for any manufacturer's innovations and powerful presentation support for any new product they accept. Services technologies have also become essential international competitive weapons for "product-oriented" companies such as Exxon or General Motors (GM). They find that information management—how well they develop and deploy worldwide information about suppliers, new technologies, exchange regulations, swap potentials, and political or market sensitivities—is as important to profits as their "production" activities.

Perhaps the most important structural change in international competition stems from the continuing integration of the world's financial centers into a single world financial marketplace. World financial flows have already become largely disconnected from trade flows.[3] Although world trade in goods and services aggregates only $3–4 trillion annually, financial transactions by CHIPS (Clearing House of International Payments) alone totaled $105 trillion in 1986, and early 1987 transactions were running more than $200 trillion on an annualized basis. Turnover on the 1986 Eurodollar market also amounted to $75 trillion. Any of these sums would dwarf world merchandise trade, just as Fedwire's $125 trillion of 1986 domestic transactions dwarfed the entire U.S. GNP. Instead of following goods, money now flows toward the highest available real interest rates or returns in safer, more stable economic situations.

Comparative costs in international competition have often become more a function of exchange rates than of productivity or competitive managerial decisions. To exploit—or avoid damage from—this situation, manufacturing strategies increasingly must be designed in a flexible three-level portfolio of manufacturing sites, sourcing locations, and geographical markets. The sophisticated global logistics and planning networks that companies have created to manage the interface among these portfolios has become a new critical competitive weapon and source of system-level economies of scale for manufacturers. At the same time, declining plant-level economies of scale in manufacturing, combined with flexible manufacturing systems, have led manufacturers to produce in smaller plants in multiple locations.

Corporate managers and public policymakers must also learn to cope with the way global integration has affected national capital markets. As was recently demonstrated, the short-run volatility of securities markets has increased enormously because of the capacity of large investor groups to move rapidly into or out of individual securities markets. The stimuli that disrupt a nation's capital markets and relative cost structures can easily come from outside sources. Consequently, it has become increasingly difficult for sovereign nations to control their economies in the short run or to fine-tune them through traditional fiscal or monetary interventions.

On the other hand, some believe that increased individual access to markets and the greater variety of hedge instruments available, which create a de facto synthetic currency beyond the control of any one country, may actually generate greater long-term financial stability (*The Economist*, 1988). Much research will be necessary to understand more fully the implications of financial markets with such rapid responsiveness and high levels of interactive feedback. In the meantime, it will be essential for nations to identify creative new trading rules or regulatory limits that can forestall potentially disastrous upheavals.

It is important to note, however, that with freer access to world capital markets everywhere, it is becoming very difficult for a single nation to maintain differentially low capital costs as a policy (as Japan has and the United States once did) in support of aggressive economic development or trading objectives. As capital costs among nations are leveled by globalized financial markets, countries such as the United States and Japan, with high domestic labor and materials costs, will be under ever greater pressures to move manufactures overseas. They will be left with few alternatives for maintaining high incomes and living standards, other than innovating—in both technical and structural terms—at a rate others cannot match. A key element will be innovation in services and with services enterprises. This will require major social and institutional, as well as business-technological, innovations.

Deregulation and Services Innovation

An important form of innovation in the United States has been early deregulation of many services, although Euromarkets were a direct consequence of the slow response of U.S. financial regulatory practices. If properly exploited by U.S. companies—and not by foreign services companies in the United States—deregulation can offer significant competitive advantages. For example, *The Economist* (1985) has argued that America's more timely deregulation of communications may make it impossible for European competitors to catch up in this decade. Banking and financial services deregulation has enabled U.S. financial companies to begin offering and managing a broad array of innovative and "derivative" products long before their more protected foreign counterparts could. In addition, transportation deregulation has forced new cost competitiveness, price structures, and innovations in management (such as hub and spoke operations) that could keep U.S. transportation costs below foreign nations' costs for years.

These new competitive structures in turn may force a revamping of regulation itself. In many areas, regulation is moving from the institutional level (the economic regulation of banks or airlines) to a functional level (regulating information disclosure, safety practices, environmental standards, communications interfaces, maintenance procedures, etc., across a range of industries) that could be much more productive. Ultimately this could mean a substantial restructuring (and hopefully elimination) of many regulatory agencies. Antitrust law may also be changing because of the realization that mergers that would once have been barred as anticompetitive make sense if (1) the true measure of the company's market share is global rather than national (Kirkland, 1988, p. 48) and (2) increased functional or cross-industry competition is taken into account. In many areas, deregulation and other accompanying innovations have already created a more decentralized, competitive, flexible, and lower cost services system for the United States, although sometimes at the cost of extensive consumer inconvenience.

Although they are not well reflected in current economic measurements, the effects of these and other institutional consolidations that services technologies allow on competitive industry structures can be profound. Some examples follow:

• Disintermediation in a number of areas has generated the same (or a better) level and quality of output with significant cost decreases. Direct access to financial services markets, of course, short-circuited many traditional banker, agent, and broker relationships. Easy direct ticketing by airlines, hotels, tours, and theaters has cut out other intermediaries.

• Large-scale efficiencies for small-scale operations occur as small local retailers or services units join into sophisticated integrated information and management networks with large suppliers or wholesalers (e.g., McKesson and Super Valu). The latter offer such services as helping retail customers locate and design their stores, advising them on increased value-added product mixes, managing stocking and displays in key departments, processing medical insurance claims for customers, recycling customers wastes through the network provider's own plants, handling customers' accounting and credit functions, and researching new uses for suppliers' products.

• Linking data bases to downstream services and production has been another structural change. Mead Data Central (with LEXIS, a legal research service, and NEXIS, a newspaper data base) and Lockheed (with Dialog Information Services) have found it natural to connect their electronic technologies to printers and to move downstream into publishing from their extensive data bases. American Airlines makes some $100 million per year by leasing its reservations system to travel agents. It is beginning to offer integrated travel reservation services with hotel and car rental chains through its AMRIS division. Citicorp is leasing to its customers access to its own extensive worldwide information network.

• Manufacturer-services integration for value added or strategic flexibility is increasingly common. GM has become the nation's largest private holder of consumer debt and is using the capabilities of its Acceptance Corporation (GMAC) as a key weapon in marketing. Oil companies now run extensive convenience food chains in parallel with their gasoline stations. Further, Sears owns or controls many suppliers for its branded lines and provides after-sale services through its own extensive financial, credit, and insurance networks.

• Privatization of government services and direct government competition with private companies have offered a new spectrum of opportunities and threats. On the one hand, state and local governments increasingly purchase services from private suppliers for efficiency. Also, in his chapter in this volume, Gönenç discusses the trend toward private competition in provision of public utilities and social services—activities which have traditionally been organized as national public monopolies. On the other hand, government institutions with their large capital resources and in-place customer bases are moving strongly into realms which would otherwise be handled by the private sector. Since 1983, Singapore has pressed heavily into electronic banking for which it hopes to become the leading center in Southeast Asia. The British Post Office, Britain's largest retail network, is spending $250 million to link itself to Clearing House Automated Payment System (CHAPS) and Society for World Interbank Financial Telecommunications (SWIFT) and to convert 9,000 of its 20,000 outlets into full bank branches and insurance offices.

Services technologies are thus breaking down the traditional boundaries of nations, industries, and even government versus private functions. With no element in its value chain (from raw materials to postsale services) immune to structural changes due to services technologies, each producer must constantly reassess who its true suppliers, competitors, and customers are and how each could enhance or subvert its competitive posture. International banks already often find that their compatriots may be simultaneously competitors, customers, joint venture partners, and suppliers. The same will happen elsewhere.

International Trade Implications

What are the implications of all these changes for the U.S. position in world trade? Only a few summary observations will be attempted here. First, unlike manufacturing, U.S. trade in services has exhibited a positive balance for years (see Table 6). Whereas the trade balance in merchandise has been negative in 12 of the last 25 years, in over half of these years services (including investment returns often representing technology advantages) have put the U.S. current account into the black. However, since 1981 the net balance of trade in services has become more tenuous (see Table 7). Inter-

TABLE 6   U.S. Net Trade Balances[a] ($ billion)

Date	Net goods and services balance	Merchandise balance	Services balance[b]	Net investment income
1965	8.3	5.0	0.2	5.3
1970	5.6	2.6	0.2	6.2
1975	22.8	8.9	1.8	12.8
1980	9.0	− 25.5	6.8	30.4
1982	0.3	− 36.4	8.3	128.7
1983	− 36.8	− 67.1	5.7	24.9
1984	− 94.8	− 112.1	1.1	18.5
1985	− 101.1	− 122.1	− 1.0	25.4
1986	− 125.7	− 144.3	1.5	20.8
Preliminary 1987	− 149.6	− 157.5	0.5	8.7

[a]Excludes military transactions.
[b]Calculated as net services transactions minus net investment income plus net military transactions.
SOURCE: Bureau of Economic Analysis, U.S. International Transactions (various years and quarters).

national Monetary Fund data show the U.S. share of world services exports dropping from 23.8 percent in 1970 to 19.2 percent in 1985.

Although trade in services is excruciatingly difficult to measure accurately,[4] most experts believe that the total volume of services trade has been seriously underestimated. U.S. government trade data track only about 40 services categories, as opposed to some 10,000 items in merchandise trade, and do not capture many important services exports. The Office of Technology Assessment (1986) estimates 1984 U.S. services exports to be $69–91 billion (as opposed to the official $43.8 billion) and U.S. imports to be $57–74 billion (versus the official $41.5 billion). Data problems are so severe that several groups in Washington are working feverishly to improve the available data for the next (Uruguay) round of General Agreement on Tariffs and Trade (GATT) discussions, where for the first time services will be a major agenda item.

One of the peculiarities of services trade is that most of the facilities and jobs created by services exports are in the user country. Unlike manufactures, relatively few services are produced in the parent country and sold for their full product value across borders. Thus, in contrast to manufacturing exports, services trade data frequently recognize only the fees or profit margins that services companies can repatriate—a small fraction of their total transactions' value.

In part because of such measurement biases, the volume of services transactions would have to expand enormously to eliminate current balance of payments deficits due to merchandise trade. Although the United States is

TABLE 7   U.S. Services Trade Balances 1982–1987 ($ million)

	1982	1983	1984	1985	1986	Projected 1987
Selected business services, total	9,212	6,585	1,957	(148)	2,559	599
Travel	(1)	(2,148)	(4,096)	(4,807)	(4,714)	(5,884)
Passenger fares	(1,598)	(2,447)	(3,474)	(4,273)	(3,280)	(3,476)
Other transportation	607	368	(1,034)	(1,786)	(1,909)	(2,530)
Fees and royalties, total	4,558	4,556	4,689	5,274	5,785	7,222
Affiliated foreigners	3,181	3,192	3,324	3,757	4,099	5,437
Unaffiliated foreigners	1,377	1,364	1,365	1,517	1,686	1,785
Other private services, total	5,646	6,256	5,872	5,444	6,677	5,267
Affiliated foreigners	2,219	3,003	2,961	3,212	4,408	2,691
Unaffiliated foreigners	3,427	3,253	2,911	2,232	2,269	2,575
Selected income transactions, total	42,833	37,868	33,031	41,201	37,130	26,807
Direct investment	18,226	14,901	11,988	26,586	30,851	28,104
Other private investment	24,607	22,967	21,043	14,615	6,279	(1,297)
Selected U.S. government transactions	(15,325)	(14,137)	(17,302)	(19,999)	(21,042)	(19,912)
Services, income and government	36,720	30,316	17,686	21,054	18,647	7,493

SOURCE: Bureau of Economic Analysis, U.S. International Transactions (various years and quarters).

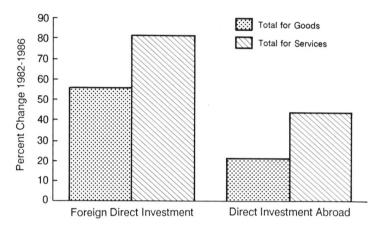

FIGURE 9 Percentage change in foreign direct investment (FDI) versus direct investment abroad (DIA): 1982–1986. SOURCE: Bureau of Economic Analysis, U.S. Direct Investment Abroad: Detail for Position and Balance of Payment Flows; Foreign Direct Investment in the United States: Detail for Position and Balance of Payment Flows.

the largest single services exporter, the European Economic Community (EEC) is larger in total volume. Proposed elimination of trade barriers within the EEC by 1992 is already spurring the development of larger and possibly stronger services sector competitors in Europe. The U.S. National Study on Trade in Services (Office of the U.S. Trade Representative, 1983) indicated that at about 1.4 percent, the United States was nowhere near the top in its percentage of gross domestic product exported as services. Expanding its services exports to match France's 5.1 percent or Britain's 6.5 percent of GNP would increase U.S. services exports by some $60–110 billion, enough to offset but not eliminate the 1986 merchandise deficit of $144 billion. Unfortunately, such gains will be hard to achieve. Countries competing with the United States are likely to become more aggressive in their own trade policies as they realize the importance of services in their own trade postures.

The overall trade situation is ambiguous. In areas such as public accounting, law, communications, and international finance, the United States enjoys a very strong position. In others such as international air travel, America's once dominant carriers Pan Am and TWA have fared badly, while competitors such as JAL, Swissair, and Singapore Airlines have thrived. The causes of such shifts are always complicated, but much of the credit for the latter's gains must go to such factors as strong government support, their heavy long-term equipment investments, and their exceptional attention to the quality of customer care on flights.

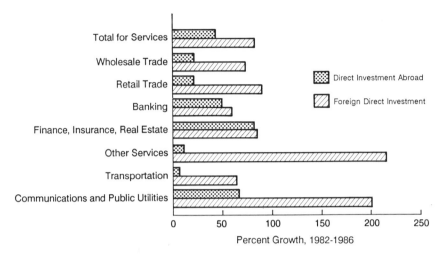

FIGURE 10   Percentage growth in direct investment abroad (DIA) versus foreign direct investment (FDI) for various services: 1982–1986. SOURCE: Bureau of Economic Analysis, U.S. Direct Investment Abroad: Detail for Position and Balance of Payment Flows; Foreign Direct Investment in the United States: Detail for Position and Balance of Payment Flows.

Unfortunately, foreign incursions in services are not limited to the international arena. Total foreign direct investment in the United States has grown by 67.9 percent since 1982, while U.S. direct investment abroad has grown by only 25.1 percent (see Figure 9). Foreign direct ·investment in the U.S. services sector has expanded rapidly (especially in communications, finance, construction, and distribution) over the last five years (see Figure 10). Many well-known U.S. names in services, such as Saks Fifth Avenue, Twentieth Century Fox, Intercontinental Hotels, Allied Stores, A&P, and Marshall Field, now have foreign owners.

Technology has made it possible for services companies to grow to a scale where they are attractive acquisitions for foreign competitors, for their even larger acquirers to manage worldwide services networks from their home bases, and for services to become a major target area for countries (such as Japan or Singapore) seeking to expand their world trade positions.

WHITHER SERVICES?

At present some U.S. services companies enjoy economies of scale and scope that their international competitors cannot equal. The notable exceptions are banking (Cacace, 1986) and communications, where the Japanese have larger enterprises due to internal structural factors (in banking) and monopoly structures (in communications). Moreover, the deregulated U.S. marketplace provides its own companies a unique stimulus for innovation. If U.S. services companies move aggressively to develop their own proprietary technology systems, they can maintain a competitive edge of one to two years in most services areas. Any slowdown or delay in such innovations will be sure to attract competitive incursions in the United States and in world services markets, such as the recent Japanese banking, tourist industry, hotel, and airline expansions and the European acquisitions in U.S. distribution and tourist trade activities. As the Wright and Pauli chapter suggests, U.S. companies and policy must move aggressively and avoid the complacency, short-term financial orientation, inattention to quality, and emphasis on scale economies (rather than customers' concerns or market responsiveness) that earlier undercut U.S. manufacturing competitiveness.

Even so, there is a nagging and real worry that an economy increasingly dominated by services can irrevocably weaken the United States in world affairs. There is no doubt that loss of manufacturing beyond a certain point can decrease both U.S. flexibilities and U.S. capabilities in defense systems. However, when manufacturing is equated with economic wealth and world power, the fact that the great nations of the past were the trading nations, the educated nations, and the money centers of the world is often forgotten. Their power was often based not on manufacturing capabilities but on superior social disciplines or organizations, raw materials or agricultural capabilities, technological skills or coalition behavior in a hostile world. Has there been a fundamental change in the last 100 years? Given the great expansion, decentralization, and integration of world manufacturing capabilities, perhaps less than many may think.

Yet even military power now seems less dependent on basic production capacity than on selected intellectual and research capabilities in a few high-technology areas, supported by specialized manufacturing, transportation, and communications services systems. The commercial airlines, communications, and information industries, for example, now provide much of the basic demand and technical competency that give the United States both its manufacturing standby and its technological capabilities for defense. Despite the growth of services, it seems unlikely that total manufacturing output capacities will drop drastically lower by the year 2000. Growing services employment and investments should help provide demands for many domestic manufactures. Some innovative manufacturers are learning to compete against low-cost foreign operations. Further, the integration of services and manufacturing systems should lead to some added manufacturing in the United States. Finally, the government will probably see that specialized manufacturing capabilities exist for key strategic purposes.

Services technologies are crucial to both the future health of manufacturing and the growth and productivity of the entire economy. Many of the greatest opportunities for entrepreneurship and applications of technology over the next two decades will—as they have for the past 20 years—lie in the services sector. One should not be afraid of, or deride, a services-dominated economy. A greater fear should be that we misunderstand the services sector, under-develop or mismanage it, and overlook its great opportunities while we attempt to shore up certain troubled manufacturing areas at great national and corporate cost.

ACKNOWLEDGMENTS

The author wishes to acknowledge the very special contributions of Jordan J. Baruch and Penny C. Paquette in the development of this paper, as well as the generous support of Bankers Trust Company, Bell and Howell, The Royal Bank of Canada, Bell Atlanticom Co., Braxton Associates, and American Express.

NOTES

1. Figures are drawn from Bureau of Labor Statistics, Establishment Data Base: States and Selected Areas (November 1978, pp. 124–133; November 1987, pp. 120–137).
2. *The Economist* (1987, p. 42) in a survey of the world economy estimated that the 1983 cost of preserving a job in the U.S. auto or steel industry was between $100,000 and $120,000.
3. Bell and Kettell (1983, p.3) estimated that 95 percent of the daily volume of foreign exchange markets in 1983 was not commercial business but trading between the foreign exchange dealers of the world's international banks.
4. For example, if a U.S. bank has a loan officer in its British subsidiary who relies heavily on data or analyses generated in its New York headquarters for a deal consummated in London, is this an export, a return on investment abroad, or just a local sale of services?

REFERENCES

Barras, R. 1986. A comparison of embodied technical change in services and manufacturing industry. Applied Economics 18(September):941–958.

Bell, S., and B. Kettell. 1983. Foreign Exchange Handbook. Westport, Conn.: Quorum Books.

Blackwell, R., and W. Talarsyk. 1983. Life style retailing: Competitive strategies for the 1980s. Journal of Retailing 59(4):7–27.

Bluestone, B., and B. Harrison. 1986. The Great American Jobs Machine. Washington, D.C.: Joint Economic Committee, U.S. Congress.

Bureau of Economic Analysis, International Investment Division, Balance of Payments Division. Foreign Direct Investment in the United States: Detail for Position and Balance of Payment Flows. Series in Survey of Current Business. Washington, D.C.: U.S. Department of Commerce.

Bureau of Economic Analysis, International Investment Division, Balance of Payments Division. U.S. Direct Investment Abroad: Detail for Position and Balance of Payment Flows. Series in Survey of Current Business. Washington, D.C.: U.S. Department of Commerce.

Bureau of Economic Analysis. The National Income and Product Accounts of the United

States. Series in Survey of Current Business. Washington, D.C.: U.S. Department of Commerce.

Bureau of Economic Analysis. U.S. International Transactions. Series in Survey of Current Business. Washington, D.C.: U.S. Department of Commerce.

Bureau of Labor Statistics. Current Labor Statistics. Series in Monthly Labor Review. Washington, D.C.: U.S. Department of Labor.

Bureau of Labor Statistics. Establishment Data Base: Average Hours and Earnings of Production or Nonsupervisory Workers on Private Nonagricultural Payrolls by Major Industry. Series in Employment and Earnings. Washington, D.C.: U.S. Department of Labor.

Bureau of Labor Statistics. Establishment Data Base: Employees on Nonagricultural Payrolls by Major Industry. Series in Employment and Earnings. Washington, D.C.: U.S. Department of Labor.

Bureau of Labor Statistics. Establishment Data Base: States and Selected Areas. Series in Employment and Earnings. Washington, D.C.: U.S. Department of Labor.

Bureau of Labor Statistics. 1987. Widespread increases in industry productivity in 1986 reported by BLS. Press release (October 1). Washington, D.C.: U.S. Department of Labor.

Cacace, L.M. 1986. Japanese banks again winners in top 500 race. American Banker (July 29):1.

Chand, R. 1986. Employment during recessions: The boost from services. Canadian Business Review (Summer):37–40.

The Economist. 1985a. The other dimension: Technology and the City of London. A survey. 296(July 6):6.

The Economist. 1985b. Telecommunications: The world on the line. A survey. 297(November 23):62.

The Economist. 1987. The world economy: The limits to co-operation. A survey. 304(September 26):66.

The Economist. 1988. Get ready for the phoenix. 306(January 9):9.

Kendrick, J. 1987. Service sector productivity. Business Economics (April):18–24.

Kirkland, R., Jr. 1988. Entering a new age of boundless competition. Fortune (March 14):40–48.

Kutscher, R. 1988. Growth of services employment in the United States. In Technology in Services: Policies for Growth, Trade, and Employment, Washington, D.C.: National Academy Press.

Kutscher, R., and J. Mark. 1983. The services sector: Some common perceptions reviewed. Monthly Labor Review (April):21–24.

Levy, F. 1987. Changes in the distribution of American family incomes, 1947–1984. Science 238(May 22):923–927.

Mark, J. 1982. Measuring productivity in the services sector. Monthly Labor Review (June):3–8.

Mark, J. 1986. Problems encountered in measuring single and multifactor productivity. Monthly Labor Review (December):3–11.

Mark, J. 1988. Measuring productivity in service industries: The BLS experience. In Technology in Services: Policies for Growth, Trade, and Employment. Washington, D.C.: National Academy Press.

McMahon, P., and J. Tschetter. 1987. The declining middle class: A further analysis. Monthly Labor Review (December):22–27.

Mishel, L. 1988. Manufacturing Numbers: How Inaccurate Statistics Conceal U.S. Industrial Decline. Washington, DC: Economic Policy Institute.

Moore, G. 1987. The service industries and the business cycle. Business Economics (April):12–17.

Office of Productivity and Technology, Bureau of Labor Statistics. 1987. Unpublished computer data sheets. Washington, D.C.: U.S. Department of Labor.

Office of Technology Assessment. 1986. Trade in Services. OTA-ITE-316. Washington, D.C.: U.S. Government Printing Office (September).

Office of Technology Assessment. 1988. Technology and the American Economic Transition. Washington, D.C.: U.S. Government Printing Office.

Office of the United States Trade Representative. 1983. U.S. National Study on Trade in Services. Washington, D.C. (December).

Personick, V. 1987. Industry output and employment through the end of the century. Monthly Labor Review (September):30–45.

Porter, M. 1985. Competitive Advantage. New York: Macmillan.

Quinn, J.B. 1987. The impacts of technology in the services sector. Pp. 119–159 in Technology and Global Industry, B. Guile and H. Brooks, eds. Washington, D.C.: National Academy Press.

Roach, S. 1985. Information economy comes of age. Information Management Review (January):9–18.

Roach, S. 1987. America's Technology Dilemma: A Profile of the Information Economy. Special Economic Study. New York: Morgan Stanley (April 22).

Roach, S. 1988. Technology and the service sector: The hidden competitive challenge. In Technology in Services: Policies for Growth, Trade, and Employment. Washington, D.C.: National Academy Press.

Rosenthal, N. 1985. The shrinking middle class: Myth or reality? Monthly Labor Review (March):3–10.

Seth, J. 1983. Emerging trends for the retailing industries. Journal of Retailing 59(3):6–18.

Silvestri, G., and J. Lukasiewicz. 1987. A look at occupational employment trends to the end of the century. Monthly Labor Review (September):46–63.

Tanaka, M. 1984. Moving beyond merchandise: Japan's trade in services. Journal of Japanese Trade and Industry (4):12–15.

Thurow, L. 1987. A surge in inequality. Scientific American 256(May):30–37.

UNIPUB. 1984. Measuring Productivity: Trends and Comparisons from the First International Productivity Symposium, Tokyo, Japan, 1983. New York: UNIPUB.

Urquart, M. 1981. The services industry: Is it recession proof? Monthly Labor Review (October):12–18.

Vollmann, T., 1986. The effect of zero inventories on cost (just in time). Pp. 141–164 in Cost Accounting for the '90s: The Challenge of Technological Change. Montvale, N.J.: National Association of Accountants.

Growth of Services Employment in the United States

RONALD E. KUTSCHER

Any examination of the services sector in the U.S. economy immediately faces the definitional question—What is to be included in "services"? There is no uniformly accepted definition. In this chapter the Standard Industrial Classification (SIC) system will be used to group industries. The SIC system broadly groups industries by whether their primary production is goods or services. The goods-producing industries most often include agriculture, mining, construction, and manufacturing.

This naturally leads to the first and perhaps the broadest definition of services: i.e., all industries not producing goods, produce services. Such industries would include transportation, communications, public utilities, wholesale trade, retail trade, finance, insurance, real estate, services, and government. Private services-producing industries include all of the industries noted above except government.

A more restricted definition of services includes only the industry group called services within the broader services-producing industry designation. This group includes personal, medical, business, and professional services.

Each of these definitions is used in this chapter to examine specific aspects of employment growth in the U.S. services-producing sector, both past and prospective. Also, employment growth by occupation and by the educational requirements of these occupations is assessed. This chapter also considers why employment in the services-producing sector is growing so rapidly. Finally, in closing, the differing characteristics of the goods- versus services-producing portions of our economy are explored briefly.

EMPLOYMENT GROWTH IN SERVICES 1972–2000

First, one must look at the key changes in U.S. employment patterns for each of the major services- and goods-producing industries from 1972 to 1986. Then, by using projections of employment to 2000, published recently by the Bureau of Labor Statistics (BLS), these industries can be examined for likely future changes.[1]

Table 1 sets forth such data on both a historical and a projected basis. From these, Table 2 presents calculated past and projected shares of employment growth.

By using three different definitions, Table 2 demonstrates the significant role of the services-producing sector in employment growth in the United States. It is also interesting to compare the various periods in terms of the relative contributions of services- and goods-producing sectors. The 1979–1986 period shows a declining employment in the goods-producing industries, particularly manufacturing (see Table 1). Consequently, the services-producing sector provided more jobs than the entire net employment growth in the U.S. economy during that period. However, during the 1979–1986 period, the rate of employment growth in the overall economy was only 1.4 percent per year, about one-half the 2.6 percent annual rate from 1972 to 1979.

There are two important factors about this slowdown to keep in mind: (1) the slowdown in growth of the overall labor force during the latter period and (2) the two recessions between 1979 and 1986, accompanied by pronounced foreign trade changes in the latter period. Projections to 2000, because of modest improvements expected in foreign trade and other factors, show no further declines in goods-producing employment (although an increase in construction employment offsets a projected decline in manufacturing). Under each of the three alternative services sector definitions, the expected share of employment growth in services from 1986 to 2000 is lower than that from 1979 to 1986, but higher than that of the 1972–1979 period.

The services industry group (the most narrowly defined designation) shows striking overall contributions to employment in both the 1979–1986 period and the 1986–2000 period. More disaggregated data for this specialized group indicate that business and medical services have led the rapid increases in this SIC group's employment growth. However, because of differing rates of measured productivity growth between the goods- and services-producing sectors, the output shift toward services is far less than the shift for employment.[2]

Occupational Employment Changes, Past and Prospective

The BLS projects that the U.S. economy will generate an increase of more than 21 million jobs between 1986 and 2000. Although this is a very high

TABLE 1 Employment by Major Sector, 1972–2000

	Employment (in thousands)			Projected 2000			Change		
Sector	1972	1979	1986	Low	Moderate	High	Low	Moderate	High
TOTAL	84,549	101,353	111,623	126,432	133,030	137,533	14,809	21,407	25,910
Nonfarm wage and salary	73,514	89,481	99,044	113,554	119,156	123,013	14,510	20,112	23,969
Goods-producing (excluding agriculture)	23,668	26,463	24,681	23,148	24,678	25,906	−1,533	−3	1,225
Mining	628	958	783	672	724	779	−111	−59	−4
Construction	3,889	4,463	4,904	5,643	5,794	6,077	739	890	1,173
Manufacturing	19,151	21,042	18,994	16,833	18,160	19,050	−2,161	−834	56
Durable	11,050	12,762	11,244	9,654	10,731	11,193	−1,590	−513	−51
Nondurable	8,101	8,280	7,750	7,179	7,429	7,857	−571	−321	107
Services-producing	49,846	63,018	74,363	90,406	94,478	97,107	16,043	20,115	22,744
Transportation and public utilities	4,541	5,135	5,244	5,410	5,719	5,903	166	475	659
Wholesale trade	4,113	5,204	5,735	7,015	7,266	7,361	1,280	1,531	1,626
Retail trade	11,835	14,989	17,845	21,795	22,702	23,079	3,950	4,857	5,234
Finance, insurance, and real estate	3,907	4,975	6,297	7,508	7,917	8,159	1,211	1,620	1,862
Services	12,117	16,768	22,531	30,778	32,545	33,708	8,247	10,014	11,177
Government	13,333	15,947	16,711	17,900	18,329	18,897	1,189	1,618	2,186
Agriculture	3,523	3,401	3,252	2,784	2,917	3,009	−478	−335	−253
Private households	1,693	1,326	1,241	1,122	1,215	1,234	−119	−26	−7
Nonfarm self-employed and unpaid family workers	5,819	7,145	8,086	8,972	9,742	10,277	886	1,656	2,191

TABLE 1 (continued)

| | Annual Rate of Change | | | | | Distribution of Wage and Salary Employment (%) | | | | | |
| | | | 1986–2000 | | | | | | Projected 2000 | | |
Sector	1972–1979	1979–1986	Low	Moderate	High	1972	1979	1986	Low	Moderate	High
TOTAL	2.6	1.4	0.9	1.3	1.5	100.0	100.0	100.0	100.0	100.0	100.0
Nonfarm wage and salary	2.8	1.5	1.0	1.3	1.6	100.0	100.0	100.0	100.0	100.0	100.0
Goods-producing (excluding agriculture)	1.6	−1.0	−0.5	0.0	0.3	32.2	29.6	24.9	20.4	20.7	21.1
Mining	6.2	−2.8	−1.1	−0.6	0.0	0.9	1.1	0.8	0.6	0.6	0.6
Construction	2.0	1.4	1.0	1.2	1.5	5.3	5.0	5.0	5.0	4.9	4.9
Manufacturing	1.4	−1.4	−0.9	−0.3	0.0	26.1	23.5	19.2	14.8	15.2	15.5
Durable	2.1	−1.8	−1.1	−0.3	0.0	15.0	14.3	11.4	8.5	9.0	9.1
Nondurable	0.3	−0.9	−0.5	−0.3	0.1	11.0	9.3	7.8	6.3	6.2	6.4
Services-producing	3.4	2.4	1.4	1.7	1.9	67.8	70.4	75.1	79.6	79.3	78.9
Transportation and public utilities	1.8	0.3	0.2	0.6	0.8	6.2	5.7	5.3	4.8	4.8	4.8
Wholesale trade	3.4	1.4	1.4	1.7	1.8	5.6	5.8	5.8	6.2	6.1	6.0
Retail trade	3.4	2.5	1.4	1.7	1.9	16.1	16.8	18.0	19.2	19.1	18.8
Finance, insurance, and real estate	3.5	3.4	1.3	1.7	1.9	5.3	5.6	6.4	6.6	6.6	6.6
Services	4.8	4.3	2.3	2.7	2.9	16.5	18.7	22.7	27.1	27.3	27.4
Government	2.6	0.7	0.5	0.7	0.9	18.1	17.8	16.9	15.8	15.4	15.4
Agriculture	−0.5	−0.6	−1.1	−0.8	−0.6						
Private households	−3.4	−0.9	−0.7	−0.1	0.0						
Nonfarm self-employed and unpaid family workers	3.0	1.8	0.8	1.3	1.7						

NOTE: Excludes SIC 074,5,8 (agricultural services) and 99 (nonclassifiable establishments); therefore, not exactly comparable to data published in the BLS *Employment and Earnings* Series.

TABLE 2   Share of Employment Growth

Sector	1972–1979	1979–1986	Projected 1986–2000
Services-producing sector	82.5	118.6	100.0
Private services-producing sector	66.0	110.6	92.0
Services industry group	29.1	60.3	49.8
Goods-producing sector	17.5	−18.6	0.0

total number, it is only one-half the average annual rate of increase that occurred over the previous 14-year period, 1972–1986[3] (see Table 3). The projected change from 1986 to 2000 is more consistent with the slower growth rates of 1979–1986 than it is with those of the 1972–1979 period.

Although overall growth in employment is slowing considerably, important changes are also occurring in the composition of employment by occupation. Faster than average growth has taken place in several occupational groups and is expected to continue in the 1986–2000 era. These include the following: (1) executive, administrative, and managerial workers; (2) professional workers; (3) technicians and related support workers; (4) sales workers; and (5) services workers. Employment declines have also occurred and are projected to continue in two broad occupational groups: (1) private household workers and (2) farming, forestry, and fishing workers. Below-average growth is expected to continue in another occupational group: precision production, craft, and repair workers. Prospects for two occupational groups are expected to change in the future as compared to past periods. The operators, fabricators, and laborers occupational group declined from 1979 to 1986, after having increased from 1972 to 1979. This group is projected to increase again in 1986–2000, but by less than 3 percent. The occupational group of administrative and support workers, including clerical, increased faster than the average from 1972 to 1979, then dropped to a below-average increase from 1979 to 1986. A slower than average increase is projected for 1986–2000 for this occupational group.

Educational Content of Jobs

Much has been written indicating that the changing occupational structure of U.S. employment will require a more highly educated work force. To see if the 1986–2000 occupational projections substantiate this view, the occupational clusters discussed in the previous section were divided into three groups defined by educational requirements. Group I included the occupational clusters in which at least two-thirds of the workers in 1986 had one or more years of college. Group II included the occupational clusters in which the median years of school completed was greater than 12 and the

TABLE 3 Employment by Broad Occupational Group for 1986 and Projected for 2000, and Change in Employment for Selected Periods

Major Occupational Group	1986 (in thousands)	2000 (in thousands)	Percent Change in Employment			
			1972–1979	1979–1986	1972–1986	1986–2000
TOTAL EMPLOYMENT	111,623	133,030	20.3	10.9	33.4	19.2
Executive, administrative, and managerial workers	10,583	13,616	34.9	28.7	73.7	28.7
Professional workers	13,538	17,192	29.8	21.4	57.5	27.0
Technicians and related support workers	3,726	5,151	39.9	24.7	74.5	38.2
Sales workers	12,606	16,334	24.3	24.4	54.6	29.6
Administrative support workers, including clerical	19,851	22,109	23.5	9.5	35.2	11.4
Private household workers	981	955	−23.0	−11.5	−31.9	−2.6
Services workers, except private household workers	16,555	21,962	25.7	16.0	45.9	32.7
Precision production, craft, and repair workers	13,923	15,590	21.7	6.5	29.6	12.0
Operators, fabricators, and laborers	16,300	16,724	8.7	−9.2	−1.3	2.6
Farming, forestry, and fishing workers	3,556	3,393	−5.1	−5.6	−10.4	−4.6

NOTE: Estimates of 1986 employment, the base year for the 2000 projections, were derived primarily from data collected in the Occupational Employment Statistics (OES) surveys. The 1972–1986 rates of change and subperiods were derived from the Current Population Survey data, because comparable data were not available for 1972 and 1979 from the OES surveys.

proportion of workers with less than a high school education was relatively low. Group III included occupational clusters in which the proportion of workers having less than a high school education was relatively high. Since workers in any occupational cluster have a broad range of educational backgrounds, the three groups can only indicate educational levels of their preponderant segment of workers; there are workers in each of the three groups at each of the educational levels.

The distribution of past and projected employment by category within these clusters is shown in Table 4. These data indicate that the occupations requiring the most education, Group I, are likely to increase as a proportion

TABLE 4 Employment in Broad Occupational Clusters by Level of Educational Attainment for 1986 and Projected 2000 Moderate Alternative

	Distribution (%)	
	1986	2000
TOTAL ALL LEVELS	100.0	100.0
Level I		
TOTAL	25.1	27.3
Management and management related	9.5	10.2
Engineers, architects, and surveyors	1.4	1.5
Natural scientists and computer specialists	.7	.8
Teachers, librarians, and counselors	4.4	4.3
Health diagnosing and treating specialists	2.3	2.8
Other professional specialists	3.5	3.7
Technicians	3.3	4.0
Level II		
TOTAL	40.8	40.0
Sales workers	11.3	12.3
Administrative support, including clerical	17.8	16.7
Blue-collar-worker supervisors	1.6	1.5
Construction trades and extractive occupations	3.4	3.3
Mechanics and repairers	4.2	4.0
Precision production, and plant and systems workers	2.5	2.2
Level III		
TOTAL	34.0	32.7
Services workers	15.7	17.2
Agricultural workers	3.3	2.6
Machine setters and operators	4.5	3.6
Assemblers and other handworkers	2.4	1.9
Transportation and materials-moving workers	4.3	4.0
Helpers and laborers	3.8	3.4

NOTE: The data in Table 4 are the same as the 1986 and 2000 data in Table 3. However, the data in Table 4 are more disaggregated so comparison is not easy.

of total employment from 25.1 percent in 1986 to 27.3 percent in 2000. The other two groups in which workers have less education are projected to decline as a proportion of total employment. The proportion of total employment is expected to decline most (from 34 percent in 1986 to 32.7 percent in 2000) in Group III, which requires the least education. The "services workers" group is in Group III, the only occupational cluster in this educational attainment group with a median of more than 12 school years completed. This group is increasing as a proportion of total employment, while all other occupational clusters within Group III are declining—some by significant amounts. Conversely, in Group I all the clusters are increasing as a share of total employment, except one—teachers, librarians, and counselors—and there the decline is very small. Projections do not allow for future educational upgrading, such as that which has taken place in the past. Even with corrections for these factors, the year 2000's share of employment in the lower educational attainment group is expected to be larger than the share of employment in the highest group.

Thus, the shift in distribution of jobs by educational level could be characterized as (1) increasing the small but most highly educated job group, while (2) decreasing the larger but less educated job groups. Each of the three broad occupational clusters in Group I—with the most highly educated workers—is projected to grow more rapidly than the national average and thus increase its share of total employment. Collectively, the three broad occupational categories in Group I, which accounted for 25 percent of total employment in 1986, are expected to provide almost 40 percent of total job growth between 1986 and 2000.

By contrast, factors such as office and factory automation, changes in consumer demand, and import substitutions are expected to lead to relatively slow growth, or a decline, for occupational groups requiring less education. Occupations such as administrative support, including clerical workers; farming, forestry, and fishing workers; and "operators, fabricators, and laborers" are examples. "Services workers," a category that is expected to grow at a faster rate than total employment and account for more of the total growth in employment than any other broad occupational group, is an important exception to the general trend, since the group's educational attainment is currently not among the highest. In broad terms, therefore, the economy is expected to continue generating jobs with higher educational requirements. This is clearly an important context in which to assess future job developments. Many—but not all—of the projected future shifts are consistent with past trends in the occupational structure of U.S. employment.

Blacks, Hispanics, and Women

Job opportunities for individuals or groups of workers are determined by a number of factors relating to the job market and the particular characteristics of workers. Consequently, in developing projections of employment by industry and occupation, the BLS does not develop projections of the demographic composition of future jobs. However, data on the current demographic composition of jobs can be used in conjunction with projected changes in employment to determine the possible implications of those projections. For example, one can see how future patterns of job growth compare with the current pattern of jobs held by blacks and Hispanics.

Blacks and Hispanics constituted approximately 10 and 7 percent, respectively, of employment in 1986. Although members of these two groups were employed in virtually every occupation, they were more heavily concentrated in some occupational clusters. The occupational clusters in Table 5 are listed in decreasing order of their projected growth rates. In general, these data show that both blacks and Hispanics are found in a greater proportion of occupations projected to decline or grow more slowly than in those that are projected to increase rapidly. These occupational clusters generally require the least amount of education and training, whereas those projected to grow fastest require the most education and training. The only exception is the services workers cluster, which is growing rapidly.

Normally, occupations having the fastest growth rates will offer better opportunities for employment. For blacks and Hispanics to improve their labor market situation, they must be able to take advantage of those opportunities. Labor force projections developed by the BLS indicate that blacks and Hispanics will constitute approximately 17 and 29 percent, respectively, of the total growth in the labor force. Because the fastest growing occupations are those in which a high percentage of workers currently have postsecondary education, improvements in educational attainment are likely to be an important factor in blacks and Hispanics being able to take advantage of these areas of rapidly growing opportunity.

The relative proportion of women in the different occupational clusters varies considerably (see Tables 5 and 6). In general, however, women have relatively high proportions of employment in the faster growing occupations, with two exceptions. Women's share of employment as natural scientists and computer specialists is currently low. As engineers, architects, and surveyors, it is even lower (7 percent). Women tend to represent smaller proportions of the occupations projected to decline or grow slowly, with one major exception—the "administrative support including clerical" category. This is a slowly growing occupational group in which women currently hold more than 80 percent of the jobs.

EXPLANATIONS FOR THE RAPID GROWTH OF EMPLOYMENT IN THE SERVICES INDUSTRIES[4]

So far, this chapter has only described the services industries' rapid employment growth patterns and characterized them by industry, occupation,

TABLE 5   Projected 1986–2000 Growth Rate, and Percent of Total Employment in 1986 Accounted for by Blacks, Hispanics, and Women

Occupational Cluster	Projected Change, 1986–2000 (%)	Percentage of Total Employment in 1986		
		Blacks	Hispanics	Women
TOTAL, ALL OCCUPATIONS	19	10	7	44
Natural scientists and computer specialists	46	6	3	31
Health diagnosing and treating occupations	42	6	3	67
Technicians	38	8	4	47
Engineers, architects, and surveyors	32	4	3	7
Services workers	31	17	9	61
Marketing and sales workers	30	6	5	48
Managerial and management-related workers	29	5	4	43
Other professional workers	26	7	4	43
Construction trades and extractive workers	18	7	8	2
Teachers, librarians, and counselors	16	9	3	68
Mechanics and repairers	15	7	7	3
Administrative support, including clerical	11	11	6	80
Transportation and materials-moving workers	10	14	8	9
Helpers and laborers	6	17	11	16
Precision production, and plant and systems workers	4	9	9	23
Machine setters and operators	−4	16	13	42
Assemblers and other handwork occupations	−4	13	11	38
Agricultural, forestry, and fishing workers	−5	7	10	16

TABLE 6 Employment by Occupation in 1986 and Projected 2000 Alternatives (percent distribution)

	1986	Projected 2000 Alternatives		
		Low	Moderate	High
TOTAL, ALL OCCUPATIONS	100.0	100.0	100.0	100.0
Managerial and management-related workers	9.5	10.2	10.2	10.3
Engineers, architects, and surveyors	1.4	1.5	1.5	1.6
Natural scientists and computer specialists	0.7	0.8	0.8	0.8
Teachers, librarians, and counselors	4.4	4.4	4.3	4.3
Health diagnosing and treating specialists	2.3	2.8	2.8	2.8
Other professional specialists	3.3	3.6	3.7	3.5
Technicians	3.3	3.8	4.0	4.0
Marketing and sales workers	11.3	12.3	12.3	12.2
Administrative support, including clerical	17.8	16.6	16.7	16.6
Services workers	15.7	17.2	17.2	17.1
Agricultural, forestry, and fishery workers	3.2	2.6	2.6	2.5
Blue-collar-worker supervisors	1.6	1.5	1.5	1.5
Construction trades and extractive workers	3.6	3.4	3.3	3.4
Mechanics and repairers	4.2	4.0	4.0	4.0
Precision production, and plant and systems workers	2.8	2.1	2.2	2.2
Machine setters and operators	4.5	3.5	3.6	3.6
Assemblers and other handworkers	2.4	1.9	1.9	2.0
Transportation and materials-moving workers	4.3	4.0	4.0	4.0
Helpers and laborers	3.8	3.4	3.4	3.4

and period. This section presents several possible explanations for this employment growth. These explanations focus on the producer-services industries—a subset of the services industry group described in the first section.[5] However, this assessment may provide a proxy for a much broader review, as one possible explanation of the overall employment shifts from the goods-producing to the services-producing industries.

Factors in Growth of the Producer-Services Industries

When using a framework from input-output analysis, several factors seem to contribute importantly to the rapid growth of producer-services industries.[6]

Growth of the Gross National Product (GNP)   One obvious explanation for
the producer-services industries' growth is the total economy's growth. Between 1972 and 1985, producer-services' output (in real terms) grew approximately 6 percent per year (see Table 7). During this period, the total
economy grew 2.6 percent per year. Thus, for those 13 years at least, gross
national product (GNP) growth explains about 40 percent of the producer-
services industries' output growth. Similarly, during the 1972–1985 period,
GNP growth explains about 50 percent of the communications industry's
output growth, about 65 percent of the medical services industries' growth,
and about 90 percent of the eating and drinking industry's output growth.

Final Demand Composition   Why do some industries, particularly producer
services, grow faster than GNP? One explanation could be shifts in the
composition of final demand or in GNP that have occurred over time. An
economy that demands more personal and medical services—as well as fewer
cars and less food—generates more employment among lawyers, guards,
and computer programmers and less employment among farmers and assembly line workers.

   Over the 1972–1985 period, the composition of final demand has changed
modestly. Consumer expenditures for durable goods accounted for approximately 8 percent of total GNP in 1972 and 10 percent in 1985. Consumer
expenditures for nondurable goods accounted for approximately 26 percent
of GNP in 1972 and 24 percent in 1985. Consumer expenditures for services
accounted for 29 percent of GNP in 1972 and 32 percent in 1985. Expenditures for investment and foreign trade as a proportion of GNP increased.
Expenditures for total government declined as a proportion of GNP, although
the share devoted to defense increased.

   What would the producer-services' output growth have been if the com-

TABLE 7   Sources of Industry Output Growth, 1972–1985 (average
annual change, percent)

| | | Output Change Explained by | | |
| | | | Composition of | |
Industry	Actual Change	GNP Growth	Final Demand	Business Practices
Services-producing	2.9	2.6	0.1	0.2
Producer services	6.0	2.6	0.1	3.3
Communications	5.5	2.6	1.1	1.8
Eating and drinking	2.9	2.6	0.0	0.3
Medical services	4.0	2.6	1.4	0.0

SOURCE: Bureau of Labor Statistics.

position of final demand alone had changed while both real GNP and business practices had not changed? "Business practices" means the way in which goods and services are assembled, as measured by the coefficients in an input-output table. The difference between the estimated output growth in this calculation and the actual growth measures the effect of changing final demand composition on producer-services industries' output growth.

In this analysis, final demand included 82 consumption groups, producer durable equipment, residential and nonresidential structures, inventory changes, exports, imports, federal government defense and nondefense expenditures, and state and local government expenditures.

Two types of changes in final demand composition were included in this calculation: (1) shifts within a given demand category (e.g., between medical services and food within the personal consumption expenditures category) and (2) shifts between broad demand categories (e.g., between the investment and total personal consumption categories). The period 1972–1985 was chosen because input-output data are available only for selected years.

These calculations show that changes in final demand composition alone increased the demand for producer services by only 0.1 percent per year for the 1972–1985 period (see Table 7). Thus, the changing composition of final demand had only a slight impact on the rapid growth of the producer-services industries. This factor explained less than 2 percent of the growth, as compared with GNP growth, which accounted for more than 40 percent of the measured growth. The size of this effect varied little with the choice of years.

Differential Effects   The changes in final demand composition did affect some services industries during the 1972–1985 period. In particular, changes in final demand caused some areas, such as the medical services and communications industries, to grow faster than GNP. However, they had relatively little total impact on the overall services-producing industries.

This minimal effect (0.1 percent per year) is easily explained. First, these producer-services industries usually sell their outputs to many other industries. The distribution of such sales roughly parallels the size of the purchasing industry. Two exceptions are purchases of engineering and architectural services by the construction industry and purchases of legal services by consumers. Second, purchased producer-services activities usually account for only 3 to 7 percent of the total cost in the industries that buy them.

The effect of changing final demand composition on medical services and communications might have been anticipated. These industries sell much of their output to consumers, and consumer expenditures for medical services and for communications grew faster than GNP over the 1972–1985 period. The effect on eating and drinking industries is modest because consumer expenditures for food purchased off-premises grew at about the same rate as GNP from 1972 to 1985.

Business Practices   Changes in business practices might also explain the above-average growth of the producer-services industries. Businesses might change the inputs they require to produce their products. Companies obviously require materials inputs such as plastics, steel, aluminum, glass, and packaging materials; but they also require services inputs such as transportation, financial, communications, maintenance, and repair services. These are producer-services activities.

Business practices affecting inputs of goods and services change over time for several reasons. New technologies or innovations become available. Relative prices change among inputs, as did the dramatically rising and falling energy prices of the 1970s and 1980s. Social changes such as deregulation change the attractiveness of different supplier groups. Changes in materials inputs are relatively easy to visualize, but changes in regulations and industry structure and practices can have dramatic effects that are hard to discern because they happen over a long period.

What would producer-services' output growth have been if business practices alone had changed while both the level of GNP and the composition of final demand had not changed? Changes in business practices were estimated from the changes in input-output coefficients for 156 industries.

Changes in business practices added about 3.3 percentage points per year to output growth in producer services (see Table 7). This represents about 55 percent of the producer-services industries' growth in the 1972–1985 period. Thus, without changes in the way goods and services are produced in our economy, the output growth of the producer-services industries over the 1972–1985 period would have been reduced by more than one-half. On the other hand, such changes added very little to the output growth of some other services industries. They explained only 0.0 to 0.3 percentage point of the growth in the broad services-producing industries as a group, the medical services industry, and the eating and drinking industry. However, they did add 1.8 percent per year to the communications industry's output growth.

The Unbundling Hypothesis

Which changes in business practices were most responsible for causing output and employment in the producer-services industries to grow at above-average rates? Some argue that the employment growth of producer-services industries reflects simply the shifting of existing legal, accounting and auditing, janitorial, or clerical activities from one industry to another. The usual reference is from manufacturing companies to services firms, i.e., manufacturing companies that once provided their own producer services now purchase these activities externally. Such transfers are called unbundling.

Unbundling, as an explanation, would imply several things. First, the number of employees involved in producer-services activities within manufacturing industries would decline over time as their functions are transferred to the producer-services industries. Second, the volume of producer-services activities throughout the total economy would not increase; only their location would change. Finally, unbundling would significantly increase demand (and employment) for the producer-services industries.

There is often confusion between unbundling and increased contracting out. Unbundling implies increased contracting, but increased contracting need not imply unbundling. Unbundling, strictly speaking, implies that the volume of producer-services activities has not increased for the total economy; only the activity's location has shifted. Increased contracting out implies that manufacturing industries are purchasing more from the producer-services industries. The increased purchases could be from unbundling, from needing new or additional producer services, or from both.

Why do companies switch from in-house staffs to outside suppliers? One reason is that businesses find it cheaper to purchase producer services from outside establishments than to perform the activities with in-house staff and capital. Supplying establishments can offer lower costs or higher quality through specialization or economies of scale in providing producer services. Manufacturing companies frequently undertake such ''make or buy'' decisions about material or energy inputs used in their production processes.

Companies also unbundle to cope with fluctuating work force requirements (Henson, 1985; Mangum et al., 1985; Piore, 1986). Rather than staffing their operations with enough permanent employees for peak production loads, companies staff with just enough permanent employees to handle average production loads and hire temporary workers (or contract out their producer services) at peak production periods. Just as companies in recent years have adopted ''just-in-time'' inventory practices, they might also adopt just-in-time personnel practices to meet overhead needs.

To analyze the unbundling phenomenon, employment trends were reviewed by both industry and occupation for the 1977–1986 period. It was difficult to completely isolate the unbundling phenomenon or to control for other factors that might affect employment trends. In these analyses the BLS Occupational Employment Statistics (OES) survey was used, which collects data on occupational employment of wage and salary workers by industry in nonagricultural establishments. Manufacturing surveys conducted in 1977, 1980, 1983, and 1986 were also used.

A major new occupational classification system was introduced in the 1983 survey. Because of this, the 1977–1980 employment estimates are not directly comparable to the 1983–1986 estimates. Also, the OES survey is conducted during April, May, and June. Thus, the employment estimates are not annual averages, but estimates for selected months.

BROAD OCCUPATIONAL TRENDS, 1977–1986

Trends in the number of wage and salary workers were observed in broad occupational groups within manufacturing during 1977–1986. Employment trends represent the net impact of changes in GNP, final demand composition, business practices, and staffing patterns. (Staffing patterns are the shares of an industry's employment accounted for by particular occupations.) Unfortunately, these trends did not provide a complete resolution of the unbundling issue. However, being the longest available trends for analysis, they have provided useful insights.

For example, the number and share of jobs held by managers employed in manufacturing increased between 1977 and 1980, and again between 1983 and 1986 (see Table 8). Managerial occupations include financial, purchasing, personnel, marketing, and administrative managers. The number of manufacturing managers increased by 201,000 between 1977 and 1980, and by 131,000 between 1983 and 1986. As noted, the 1980–1983 decline resulted largely from new occupational definitions.

The share of all wage and salary jobs in manufacturing held by managers also increased: from 5.7 percent in 1977 to 6.6 percent in 1980, and from 5.8 percent in 1983 to 6.4 percent in 1986. These increasing levels and shares suggest that manufacturing industries did not unbundle managerial-type producer services during these periods.

Similar changes occurred among the professional, paraprofessional, and technical occupations within manufacturing. Included in these groups are accountants, engineers, scientists, computer scientists and programmers, and engineering and science technicians. The number of professional and technical workers increased approximately 336,000 between 1977 and 1980, and 239,000 between 1983 and 1986. The share of manufacturing employment accounted for by professional, paraprofessional, and technical occupations also increased from 8.4 percent in 1977 to 9.9 percent in 1980, and from 11.0 percent in 1983 to 11.8 percent in 1986. Such increasing levels and shares suggest that producer-services activities related to professional and technical occupations were not unbundled by manufacturers during these periods.

A different picture occurs for clerical and administrative support occupations. (Clerical workers include secretaries, computer operators, bookkeepers, and dispatching and inventory clerks.) The number of clerical workers employed in manufacturing increased by 162,000 between 1977 and 1980, and by 49,000 between 1983 and 1986. (The 1980 and 1983 estimates shown in Table 8 are not comparable because of changes in the coding structure.) The importance of clerical occupations in manufacturing appears to have peaked in the early 1980s. In 1977, wage and salary workers in clerical occupations accounted for 11.0 percent of total manufacturing employment;

TABLE 8 Employment Trends for Selected Broad Occupational Groups Within Manufacturing, 1977–1986

Occupation	Numbers (in thousands)		Percent Distribution		Numbers (in thousands)		Percent Distribution	
	1971[a]	1980[a]	1977[a]	1980[a]	1983	1986	1983	1986
TOTAL EMPLOYMENT	19,772	20,228	100.0	100.0	18,369	19,042	100.0	100.0
Managers and administrative workers	1,127	1,328	5.7	6.6	1,062	1,193	5.8	6.4
Professional, paraprofessional, and technical workers	1,662	1,998	8.4	9.9	2,013	2,252	11.0	11.8
Clerical and administrative support workers	2,160	2,322	11.1	11.5	2,151	2,200	11.7	11.6
Services occupations	390	373	2.0	1.8	326	302	1.8	1.6
Sales workers	419	439	2.1	2.2	541	611	2.9	3.2
Production and related workers[b]	13,964	13,767	70.8	68.1	12,277	12,484	66.8	65.6

NOTE: The 1986 data are unpublished and subject to revision.

[a]Because of revisions in occupational definitions introduced with the 1983 data, the 1977 and 1980 estimates are not comparable to the 1983 and 1986 estimates. For 1977 and 1980 estimates, professional and technical occupations were combined.

[b]For the 1983 and 1986 estimates, production and agricultural workers were combined.

in 1980, 11.5 percent. Between 1983 and 1986, the share of clerical occupations within manufacturing declined slightly from 11.7 to 11.6 percent. The increasing employment levels, but declining share figures, in the 1983–1986 period suggest a structural change warranting further exploration.

The pattern for services occupations (including, for example, guards and janitors) within manufacturing is also complicated. The number of services workers employed in manufacturing declined from 1977 to 1980 by 17,000, and from 1983 to 1986 by 23,000. The importance of services occupations in manufacturing thus appears to have declined since 1977. Services occupations accounted for 2 percent of manufacturing employment in 1977, and 1.8 percent in 1980; their share declined from 1.8 percent in 1983 to 1.6 percent in 1986.

Employment estimates for sales and production occupations are also shown in Table 8. However, they are not discussed here because they are seldom the focus of the unbundling argument.

More Detailed Analyses of the 1983–1986 Period

To determine the extent to which manufacturing industries unbundled clerical and services activities, the sources of such occupational changes must be isolated. If the impacts of changes in final demand composition, business practices, and labor productivity on clerical occupations within manufacturing can be estimated, one could focus clearly on unbundling.

This more detailed analysis is limited to 1983–1986 because of the change in occupational definitions introduced in the 1983 OES. This is not a problem in the analysis because only the manufacturing employment trends since the 1981–1982 recession concern most analysts. Manufacturing employment did not recover as quickly from the last recession as from previous recessions. Unbundling is one of several explanations given for the slow recovery.

One explanation for the employment growth of clerical occupations over the 1983–1986 period is the total employment growth in manufacturing industries during this era. Data from the OES show that wage and salary employment in manufacturing increased by 673,000 over the three-year span. If the percentage of clerical workers had remained at 11.7 percent of total manufacturing employment, as in 1983, employment in clerical occupations would have increased by 79,000 (see Table 9). However, the actual increase in clerical occupations within manufacturing was only 49,000. Something else caused the employment of clerical workers to lag behind the growth of total manufacturing employment over the 1983–1986 period. Repeating this analysis for other occupations, one finds that if total manufacturing employment growth were the only change between 1983 and 1986, the number of workers in managerial occupations in manufacturing would have increased by 39,000, as compared to an actual growth of 131,000. A similar analysis

TABLE 9 Sources of Occupational Change in Manufacturing
Employment, 1983–1986 (numbers in thousands)

Occupation	Actual Change	Employment Change Explained by			
		Total Manufacturing Employment Growth	Industry Mix	Composition of	
				Staffing Patterns	Other[a]
Managers and administrative workers	131	39	9	81	2
Professional, paraprofessional, and technical workers	239	74	36	118	11
Clerical and administrative support workers	49	79	17	44	− 2
Services occupations	− 23	12	− 1	33	− 1
Sales workers	70	20	5	37	2
Production and related workers	207	450	− 66	166	− 12

[a]Residual effects.

reveals that the number of workers in professional, paraprofessional, and technical occupations would have grown by 74,000 rather than the actual growth of more than 239,000. Finally, the number of workers in "services occupations" within manufacturing would have increased by 12,000 if the only change were in the level of manufacturing employment. Employment in such occupations actually declined by 23,000. Again, another factor was causing employment of services workers to lag behind total employment growth in manufacturing.

Industrial Composition, 1983–1986

Another potential source of employment growth among clerical occupations could be the changing mix of manufacturing industries. Here, industry mix means the number of workers employed in a particular industry as a percentage of total manufacturing employment. For example, the motor vehicles industry accounted for 4.1 percent of the manufacturing workers in 1983 and 4.6 percent in 1986. The construction machinery industry accounted for 1.4 percent of all manufacturing workers in 1983 and 1.2 percent in 1986.

What would the change in clerical employment between 1983 and 1986 have been if industry mix alone had changed while both the level of manufacturing employment and the proportion of clerical workers within the individual manufacturing industries had not changed? The difference between this employment estimate and the number of clerical workers in total manufacturing in 1983 measures the impact of the changing mix of manufacturing industries. This analysis uses employment trends for 143 manufacturing industries to measure the changes in industry mix. The industries are at the three-digit level of the SIC.

The changes in industry mix themselves are the results of other forces including productivity trends among the various industries, changes in the composition of final demand, and changes in business practices. Both the level of total manufacturing employment and changes in staffing patterns among the individual industries are held constant in this step.

Analysis on this basis shows that the changing mix of manufacturing industries alone caused the number of clerical workers in total manufacturing to increase 17,000 between 1983 and 1986. It also explains 9,000 of the actual 131,000 increase in the number of managers and 36,000 of the 239,000 increase in the number of professional, paraprofessional, and technical workers. Industry mix changes alone would have caused employment among services workers to decline by 1,000.

Industry Staffing Patterns, 1983–1986

A final potential explanation for the employment growth of clerical workers in manufacturing from 1983 to 1986 could be the changed staffing patterns among the detailed manufacturing industries. Staffing patterns are the proportion of employment accounted for by clerical occupations within a particular industry. What would the change in occupational employment have been if staffing patterns alone had changed while both total manufacturing employment and the composition of industries had not changed? Holding these two elements constant implicitly holds final demand composition, business practices, and industry productivity constant between 1983 and 1986. The difference between this estimated employment and the actual 1983 employment isolates the effect of changing staffing patterns.

If staffing patterns among the detailed manufacturing industries were the only change for this period, total employment of clerical workers within manufacturing would have declined by 44,000 (see Table 9). On this basis most manufacturing industries employed proportionately fewer clerical workers in 1986 than in 1983. However, for clerical occupations, the growth in total manufacturing employment more than offset changes in staffing patterns within the detailed industries—resulting in the observed net increase of 49,000 clerical workers over the 1983–1986 period.

Similarly, if staffing patterns were the only change in 1983–1986, employment in (1) managerial and (2) professional, paraprofessional, and technical occupations would have grown by 81,000 and 118,000, respectively. However, the data show that manufacturing industries employed proportionately more workers in these occupations over the three years.

Combining the Three Factors

The three employment estimates can now be combined to help understand the changes occurring in clerical employment within manufacturing. The number of clerical workers would have increased by 79,000, based on total manufacturing employment growth alone. The number would have increased 17,000, based on changing industry mix alone. The number would have declined 44,000, based on changes in industry staffing patterns alone. As noted, the actual change was 49,000. The decline isolated by changing staffing patterns alone was the only estimate that indicates potential unbundling.

Similar analyses of services workers employed in manufacturing indicate that some unbundling could be occurring. The number of services workers did decline by 23,000 between 1983 and 1986. The changes in staffing patterns among the detailed industries alone would have caused a decline of 33,000; however, this effect was reduced by the total employment change in manufacturing, whereas the effects of changing industry mix were slightly negative (1,000 jobs).

Estimates for professional, paraprofessional, and technical occupations yield a different picture. The three effects were all positive. Changes in the level of manufacturing employment alone explained 44,000 of the actual 239,000 increase in the number of these workers; changes in industry mix alone explained 36,000; and changes in staffing patterns alone explained 118,000. From these three positive effects, one can conclude that unbundling of professional-type activities did not occur.

Impact on Producer-Services Industries

What does the decline of 44,000 in employment among clerical occupations that can be explained by changing staffing patterns mean? One possibility is potential unbundling. Individual manufacturing industries might be employing proportionately fewer clerical workers because of unbundling. However, available data do not permit isolation of the causes of changing staffing patterns; the estimate represents the net effect of many other factors, such as technology and business cycle shifts, as well as potential unbundling. If unbundling were the sole explanation for the changes in staffing patterns, the estimate of 44,000 would be equivalent to about 2.8 percent of the total employment growth of producer services. The total number of workers in producer services increased 1,554,000 between 1983 and 1986. For the 44,000 estimate to be a result of unbundling would require all the producer-services activities related to these jobs to be simply transferred from manufacturing industries.

Although we do not know whether the estimate of 44,000 was a direct transfer of activities from manufacturing to producer services, we do know that unbundling did not alter the staffing patterns in the producer-services industries. According to data from the Current Population Survey (CPS), the proportion of producer-services employment accounted for by clerical workers changed very little between 1983 and 1986.[7] This demonstrates strongly that all activities within producer services grew, not just clerical-type activities.

The employment of services workers within manufacturing declined 33,000 because of changes in staffing patterns alone. If these workers were reabsorbed in producer services, this estimate would account for 2.1 percent of the total employment growth.

According to the above calculations, unbundling is not even a potential explanation for the trends in managerial, professional, paraprofessional, and technical occupations within the producer-services industries. Changes in staffing patterns alone caused employment in these occupations to increase within manufacturing industries.

Conclusions About Unbundling

From the above evidence it can be concluded that unbundling has been a very small factor in the employment growth of producer services. Occupational employment trends within manufacturing show that unbundling is not occurring for the broad managerial, professional, and technical occupations within manufacturing. Relative employment in these occupations is actually increasing. Unbundling is potentially a factor that explains some trends affecting clerical and services occupations within manufacturing. However, employment shifts due to changing staffing patterns could account for only a small proportion of the total employment growth in producer services. Unbundling could, of course, be occurring in individual firms, because trends noted here were the net effects of all the individual firms at the industry or sector level. Thus, considerable unbundling at one firm could have been offset by employment growth in the same occupations at other firms.

WHY DO BUSINESSES DEMAND MORE PRODUCER SERVICES?

Why have businesses demanded more producer-services' inputs to make their products? Why has contracting out increased? Changing business practices explain a large proportion of the producer-services sector's output growth.

Since these changes are not due to unbundling, the increased contracting must be for new or additional services. The remainder of this section lists some suggested explanations for increased contracting out, but the merits of each explanation are not reviewed.

Economies of Scale in Information The growth of employment in producer services may be a response to increasing demands for information (Browne, 1986). Computer and data processing technologies allow services units to spread the high investment and development costs of their systems over many users, thus providing economies of scale. Similarly, management and business consulting services, engineering and architectural services, and other producer services have spread the costs of acquiring sophisticated technical knowledge about demographics, economics, marketing, engineering, and other fields over many users.

Specialized Corporate Services The increasing number of large companies and conglomerates may have created a demand for specialized producer services (Cocheba et al., 1986; Stanbeck, 1979). According to this argument, a typical 1980s corporation is often diversified into many fields or industries, including manufacturing, retail, transportation, or personal services activities. A 1960s corporation might have been involved in fewer fields or industries. Thus, top managers in today's more complex enterprises must increasingly rely on experts in sophisticated producer-services units (such as business management and consulting firms) to ensure efficient operations.

Government Regulations and Laws Some argue that there are more lawyers, accountants, and other technical experts today than in the 1950s and 1960s because of the number and complexity of laws passed by Congress, state legislatures, and city councils.[8] New regulations and laws often present specialized problems in dealing with banking, construction, environment, labor relations, safety, transportation, and other services issues.

International Trade The growth of producer-services industries may be explained in part by the fact that such services have reached a scale and complexity where they can be exported themselves (U.S. Congress, Office of Technology Assessment, 1986).

Resistance to Innovation One argument is that services-producing industries and workers resist innovations over time (Bailey, 1986; Baumol, 1967; Baumol et al., 1985; Thurow, 1981, 1985). According to this explanation, the economy can be divided into two industries: stagnant and progressive. Stagnant industries resist innovations; progressive industries rapidly incorporate change. If relative demands were unchanged over time, the stagnant industries would slowly absorb more and more of the economy's inputs. If this argument is valid, lawyers (and legal firms), janitors (and services to dwellings and other buildings), guards, accountants, and computer programmers should be performing their tasks the same way they did 10 to 20 years ago.

Durability and Commonality   One suggestion is that services are expanding because there is little opportunity to lower unit costs through repeating production. For example, one transaction often has little in common with the next in legal, automobile repair, or banking services.[9] In contrast, the production of goods such as automobiles or wheat is more likely to be replicable in large quantity, thus allowing capital substitution and experience curve effects to reduce unit costs.

Finally, there are serious problems of data definitions. How does one define output units and quality in services? Are goods really more durable and portable than services, thus creating greater long-term value in their customers' hands? Computer software programs are both durable and portable. Some producer services such as temporary help or janitorial services are neither durable nor portable. Legal, engineering, or management consulting is highly portable by telecommunications and air travel. The benefit of a lawyer's or an engineer's work is as durable as the output of most manufacturing industries.

WHY THE CONCERN ABOUT A SERVICES ECONOMY?

Frequently one hears the argument that our economy cannot endure as just a services-producing society or that it cannot exist if people merely shine each other's shoes or take in each other's laundry. The reasoning for this concern, although never explicitly stated, seems to be as follows: (1) the production of goods is better because it generates more income than the production of services; (2) the production of services can take place only after the demands for goods have been satisfied; and (3) the country cannot maintain its wealth—or indeed survive—by producing only services. Why is the production of goods viewed as good, whereas the production of services is viewed as bad, poor, or at least less good?

The argument that the production of goods is more income-generating than the production of services implies that the production of goods valued at one dollar generates more useful income than the production of services valued at the same price. Is the economy better off because a consumer spends $100 on skis rather than spending $100 on ski lift tickets—the former being goods, the latter services? This seems doubtful. If consumers spend their own dollars to maximize marginal gains to themselves, the effect of the $100 depends not so much on what the consumer spends it for, but on how many times it is respent. On average in our economy, about two-thirds of spending ultimately translates into wages and benefits payments to wage earners. This average, however, varies considerably by industry.

One basis for the argument that the production of goods is better than the production of services may rely on the assumption that goods last longer than services and, therefore, can be used again and again. Certainly, that is true of some goods, but other items currently classified as the products of a goods-producing industry (such as orchids, newspapers, or beer) are very short-lived. The contrary assumption is that a service is something we consume immediately which, therefore, is not long-lasting. That certainly is not true for the products of services-producing industries such as life insurance, computer programming, estate planning, or funeral services—the latter being the most lasting of all.

Problems of Classification

It is not always easy to determine why an item is classified as it is in the industrial classification system currently in use. Many times the groupings are relatively arbitrary. Newspapers are classified in manufacturing, a goods-producing industry, whereas computer programming is often classified as a services-producing industry. If, in the production of goods, accountants or computer programmers are directly on the payroll of a manufacturing firm, their work is classified as contributing to the production of goods. However, if the accountants' or computer programmers' services were contracted for by the same manufacturing firm, they would be classified as services. The same holds for other activities that can be performed within a firm or contracted out, such as legal services, word processing, or building maintenance, to mention only a few.

Perhaps the notion that goods production is more income-generating than services production may be linked to the presumption that investment—an activity closely associated with the accumulation of some types of goods—may be more wealth-enhancing than current consumption. This would generally be true of buildings, machine tools, or construction machinery—investment goods that can be more wealth-enhancing to the economy in the long run than can the production of many specific services. However, many other goods (e.g., processed foods, soft drinks, clothing, newspapers, and magazines) are current consumption items, not contributing much to wealth accumulation. By contrast, education—the product of a services-producing industry—enjoys perhaps the greatest long-term, wealth-producing potential of any single industry.

The notion that goods production is of higher value than the production of services may also be based on a confusion over end products. Some suggest that services can only achieve permanent value by becoming inputs to goods production. Yet goods themselves can be either end products (such as refrigerators and automobiles) or inputs to later stages of production (such as plywood and sheet steel). Similarly, services such as advertising can be inputs (to either goods production or the production of other services), or they can be final products used in the form produced (such as recreation—a round of golf or an opera performance).

Services—A Prime Mover or a Dependent?

Can a service be a prime economic mover and exist without prior goods production? Or must an economy have goods production first in order to support the services that follow? There is no reason why medical, legal, entertainment, education, banking, or many other services should require prior goods production to be effective. Such services activities require only that there be income for individuals to buy the services. The source of that income can be employment in either a goods- or a services-producing industry, or other income such as rent, interest, savings, or dividends. Goods-producing industries (such as household furniture) likewise require only that individual buyers have the necessary income, regardless of source; hence, no differentiation appears.

What about services-producing industries that primarily serve other businesses, as advertising does? If the industry served is a goods-producing industry, the goods-producing industry is clearly the prime mover. However, many goods-producing industries manufacture items, such as logs, fasteners, steel, newsprint, or chemical feedstocks, that are inputs to other goods- or services-producing industries and thus depend equally on those other industries. By contrast, hospitals, which require some inputs (such as medical instruments) from goods-producing industries, are also really economic prime movers, even though they are classified as a services-producing industry. Again it can be argued that no meaningful differentiation exists in an economic sense.

The prime mover notion may be related to another argument: that goods production requires more primary inputs and produces more repair and maintenance tasks later, whereas services do not create such secondary consequences. Closer examination of goods-producing and services-producing industries reveals that although some goods (such as automobiles) do require many inputs and future servicing activities, this is not true of many other goods (such as books or magazines). Some services (electricity generation or air transportation) are very complex and require many inputs as well as processing, monitoring, and later maintenance. Again, the differentiation between goods-producing and services-producing industries seems arbitrary.

Can the country survive producing only services? There is no basis for the belief that goods production is more necessary to an economy, more income-generating, or more wealth-enhancing than services. In the past it may have been thought that only goods could enter into foreign commerce, but today services do also. Although goods production is useful, desirable, and even vital from a national defense viewpoint, it does not follow that the economy exists only because of goods production. An economy could survive, at least in theory, producing only services—if that economy's competitive advantage was in the production of services. This may be the current position of the United States with respect to a number of services and one of the factors contributing to the very rapid growth of employment in some of the services-producing industries in this country.

Thus, the arguments typically advanced for considering services industries as inferior to goods-producing industries seem not to have much validity. An economy with a mixture of goods and services production is unquestionably stronger. The current mix of goods and services in our economy is a response to what purchasers are demanding and not a cause for the alarm it seems to generate. There is no good argument for advocating a stronger shift to services-producing industries or for retarding that shift in some arbitrary way.

SUMMARY

Services have contributed significantly to job growth in the U.S. economy and are likely to continue to do so. The growth of the services economy should, however, not be cause for widespread concern. It is largely the growth of demand and employment in the services-producing industries—and lesser demand and growth for U.S. goods-producing industries—that has caused the large relative shift toward services. One element of this shift is the real growth in producer services.

There are several explanations for the above-average growth of the producer-services industries. The most important explanation for this growth is a structural change in the way our economy produces goods and services. Based on the available evidence, the unbundling of services activities from manufacturing operations accounted for only a small portion of the total producer-services industries' above-average growth. However, unbundling in individual firms could have been offset by employment growth for these same activities in other firms. Also, in individual unbundling situations, there may be displacement—as opposed to transfer—of individual workers. The impact of such unbundling on individuals must always be recognized.

Finally, profound implications result from these and other trends in the type of employment and educational preparation required. Job requirements are moving over the long run toward those jobs with higher educational requirements. All factors point toward the continuing need for a better educated work force. Our data strongly indicate that those who are not undertaking such an education will be increasingly at risk in the future job marketplace.

NOTES

1. Projections of employment for 2000 are taken from Personick (1987).
2. This point is explored further in Kutscher and Personick (1986, pp. 3–13).
3. Occupational projections are taken from Silvestri and Lukasiewicz (1987).
4. This section draws very heavily on Tschetter (1987).
5. The producer-services industries are part of the services industry grouping (narrowly defined). This encompasses advertising, computer and data processing services, personnel supply services, management and business consulting services, protective and detective services, services to dwellings and other buildings, legal services, accounting and auditing services, and engineering and architectural services.

    This section of the chapter uses numerous data sources. When describing employment trends in the nonagricultural industries and producer services, the Current Employment Statistics (CES) are used. When describing occupational employment for manufacturing industries, the Occupational Employment Survey (OES) is used.

    An alternative industry data source is the BLS Current Population Survey (CPS), which is compiled from household interviews. There are important differences among the surveys. The CPS counts the number of persons who are employed; the CES and OES count jobs. Because of this difference, a person holding two or more jobs would be counted two or more times in the CES and OES, but only once in the CPS. Another difference is that the CPS includes estimates of self-employed workers, unpaid family workers, and wage and salary workers; the CES and OES include only wage and salary workers.
6. When viewed in a descriptive sense, as a system of data classification and accounting, input-output is generally acceptable to all economists. However, here input-output is used as a theory of production, with the assumption that the coefficients comprise a set of technological parameters in a linear homogeneous production function with fixed proportions among the various inputs. For another example of the analysis used in this chapter, see General Accounting Office (1987).

    The input-output tables used in this chapter are developed by the BLS and based on tables prepared by the U.S. Department of Commerce, Bureau of Economic Analysis. For a description of input-output tables, see U.S. Department of Commerce (1984). In the Department of Commerce's input-output tables, industrial purchases of producer services are usually based on occupational employment patterns: total receipts of a particular service are usually distributed to the purchasing industry based on the number of persons in a particular occupation in the purchasing industry. See U.S. Department of Commerce, Bureau of Economic Analysis (1980).
7. The analysis presented in this section was repeated with CPS data. The results from the CPS analysis were virtually identical, i.e., some unbundling might be occurring but it would explain very little of the employment growth in the producer-services industries. The analysis presented in this chapter is better because the OES is based on a substantially larger sample.
8. There is no specific proponent of this explanation, but it is reviewed in many studies of producer services. See McCrackin (1985).
9. For a discussion of the Producer Price Index and issues concerning price indices, see Bureau of Labor Statistics (1982).

REFERENCES

Bailey, M. N. 1986. What has happened to productivity growth? Science (October):443–452.

Baumol, W. J. 1967. Macroeconomics of unbalanced growth. American Economic Review (June):415–426.

Baumol, W. J., S. A. B. Blackman, and E. N. Wolff. 1985. Unbalanced growth revisited: Asymptotic stagnancy and new evidence. American Economic Review (September):806–817.

Browne, L. E., 1986. Taking in each other's laundry: The service economy. New England Economic Review (July/August):20–31.

Bureau of Labor Statistics. U.S. Dept. of Labor. 1982. BLS Handbook of Methods. (Bulletin 2134-1) Washington, D.C.: U.S. Government Printing Office.

Cocheba, D. J., R. W. Gilmer, and R. S. Mack. 1986. Causes and consequences of slow growth in the Tennessee Valley's service sector. Growth and Change (January):51–65.

General Accounting Office. 1987. Bureau of Labor Statistics employment projections: Detailed analysis of selected occupations and industries. Report GAO/OCE-85-1 (April 15).

Henson, R. C. 1985. Coping with fluctuating work-force requirements. Employment Relations Today (Summer):149–156.

Kutscher, R., and V. A. Personick. 1986. Deindustrialization and the shift to services. Monthly Labor Review (June):3–13.

Mangum, G., D. Mayall, and K. Nelson. 1985. The temporary help market: A response to the dual internal labor market. Industrial and Labor Relations Review 38(4):599–611.

McCrackin, B. H. 1985. Why are business and professional services growing so rapidly? Economic Review (Federal Reserve Bank of Atlanta) (August):15–28.

Personick, V. A. 1987. Industry output and employment through the end of the century. Monthly Labor Review (September):30–45.

Piore, M. J. 1986. Perspectives on labor market flexibility. Industrial Relations XXV(2):146–166.

Silvestri, G. T., and J. M. Lukasiewicz. 1987. Projections 2000: A look at occupational employment trends to the year 2000. Monthly Labor Review (September):46–63.

Stanbeck, T. M., Jr. 1979. Understanding the Service Economy. Baltimore: Johns Hopkins University Press.

Thurow, L. C. 1981. Strengthening the Economy. Washington, D.C.: Center for Democratic Policy.

Thurow, L. C. 1985. Pruning our white-collar ranks: A key to productivity. Technology Review (November/December):14–15.

Tschetter, J. 1987. Producer services industries: Why are they growing so rapidly. Monthly Labor Review (December):31–40.

U.S. Congress, Office of Technology Assessment. 1986. Trade in Services: Exports and Foreign Revenues—Special Report. OTA-ITE-316. Washington, D.C. (September).

U.S. Department of Commerce. 1984. The input-output structure of the U.S. economy, 1977. Survey of Current Business (May):42–79.

U.S. Department of Commerce, Bureau of Economic Analysis. 1980. Definitions and Conventions of the 1972 Input-Output Study (July 4). Washington, D.C.: U.S. Government Printing Office.

Role of Services in the U.S. Economy

FAYE DUCHIN

Finally, we call "tertiary" products or services like the artist's tapestry, justice, education, or a man's haircut which have the two properties of (1) undergoing very little technical progress and (2) facing steeply increasing consumer demand, with no sign of its leveling off in any country.

J. FOURASTIE

Strategies for improving the performance of the U.S. economy sometimes accord particular importance to the distinction between manufacturing activities and the production of services. One view is that the continued growth of services sectors represents a natural progression out of manufacturing into the postindustrial age run by computer-expert information workers. According to this view, as other countries take over the dirty work of manufacturing, the U.S. needs to create markets for information-based services and act energetically to overcome formidable barriers to their international exchange so that the U.S. can specialize in their production and distribution.

A different perspective is that a strong manufacturing base is indispensable for the nation as a whole and that its neglect has been a leading cause of present economic problems. Proponents are likely to favor investment in manufacturing research and development, modernization of production facilities, management reorganization, and other ways to reduce the cost and improve the quality of U.S. manufactured goods. Although they generally recognize the importance of also investing in at least a small set of "high-tech" services, they might characterize most services as intrinsically labor-intensive consumer items (such as concerts and haircuts) or as representing a changing division of labor ("taking in each other's wash") that does little to augment national wealth.

That both of these views are plausible is explained in large part by the fact that the services sectors, which today employ over 70 percent of the U.S. labor force, are extremely diverse. The distinction between all manufacturing on the one hand and all services on the other cannot provide even a general basis for operational industrial and trade strategies. In fact, it is not the distinction between manufacturing and services but their detailed interdependence that is investigated in this chaper.

In the following sections, the nature of services activities is examined. Then a quantitative analysis of the relationships among services and manufacturing sectors is presented, followed, in conclusion, by some policy and strategy implications of the analysis.

THE NATURE OF SERVICES

The work of artists and barbers—not to mention magicians, athletes, clergy, and chiropractors—is often evoked, even by technical analysts, to distinguish the delivery of services from the production of physical goods. Such workers symbolize the individual creativity, judgment, skill, and physical presence for which machines presumably cannot substitute. Furthermore, the outcome of much of this labor provides uplift and healing, the demand for which is likely to expand rather than diminish as people's material needs are increasingly satisfied. These attributes are associated with the services sectors.

The computer's ability to play a role in the production of art and music has already been demonstrated; and various recording, transmitting, and playback technologies enable a single physical performance—cultural, athletic, and most recently, religious—to be experienced repeatedly by audiences of unlimited size around the globe. These technologies will become increasingly refined and increasingly prevalent as will, for example, medical diagnostic expert systems to supplement the judgment of even the most distinguished specialists. Although today's sheepshearing robots are not promising as prototypes for the displacement of barbers, it is clear that new technology transforms not only routine but also highly individualistic work.

Besides the fact that their work will be affected by new technologies, artists, barbers, and the others are hardly typical services sector employees. They account for only a small portion of the labor force—even the services sector labor force—most of whom work in offices or stores with wordprocessors, telephones, computers, merchandise, and customers, engaging in relatively structured transactions. The importance of the image of the artist or the barber is not that it is a realistic portrayal of work in services sectors but rather that it succeeds in isolating those functions that are the poorest candidates for routinization, and therefore for automation, and in associating these abstract functions in an intuitively convincing way with concrete, occupational categories (e.g., creativity with the artist, a simple but unstructured task environment with the barber). It is the nature of automation that an increasing portion of work time will be devoted to such functions, in many manufacturing and services sector jobs, as the more routine work functions are automated.

The other colorful characterization of services work, taking in each other's wash, alerts us to the fundamental importance of the changing division of labor between household and marketplace. Has the number of services jobs increased because women enter the labor force mainly to work in laundries and fast-food restaurants and to drive school buses, instead of doing the washing and cooking at home and chauffeuring their own children? This is surely true to some degree, just as businesses now buy certain services they would once have performed for themselves. The extent of this phenomenon and its implications have never been systematically analyzed. This important question is addressed later in the context of constructing an appropriate conceptual framework.

THE USE OF SERVICES IN PRODUCTION AND CONSUMPTION

The diversity of services is illustrated by the 18 services sectors identified in Table 1, which shows the proportion of each sector's output absorbed in production.[1] The remaining proportion, not used in production, is delivered mainly to public and private consumption.[2] Three categories of services can be distinguished: those used mainly in production (I), those delivered about equally to production and for public and private consumption (II), and the remainder (III) serving primaily final consumers. Four of the services sell primarily to businesses: Radio and TV Broadcasting, Business Services, Utilities, and Transportation and Warehousing; this concentration is projected to continue to increase for Business Services and Utilities by 1990.

Ten more sectors deliver mainly to consumers, especially Private Education and Health and Hospital Services that are used almost exclusively by individuals in their personal capacities. The remaining five sectors serve production and consumption about evenly: Wholesale Trade, Insurance, Communications, Government Postal Services, and Finance. The size and importance of these five sectors confounds any simple dichotomy between producers' and consumers' services.

Each of these still broadly defined sectors includes numerous individual services, and a more detailed breakdown for Business Services and for Transportation and Warehousing is illustrated in Table 2. Of the 21 subcategories identified, all but four still delivered more than 60 percent of their output for use in production.

TABLE 1 Percentage of Services Sector Outputs Absorbed in Production in Benchmark Years from 1963 to 1990

Category	Sector[a]		1963	1967	1972	1977	1990[b]
I	69	Radio/TV Broadcasting	97	99	97	97	97
	77	Business Services[c]	76	79	81	81	92
	70	Utilities	61	62	63	68	71
	67	Transportation; Warehousing Services	66	65	64	64	63
II	71	Wholesale Trade	52	51	54	55	56
	74	Insurance	53	50	49	53	52
	68	Communications[c] (except no. 69)	55	55	55	48	46
			55		54	48	51
	85	Government Postal Services		52			
			40	37	44	45	42
	73	Finance[c]					
III	79	Auto Repair and Services	37	43	43	41	44
	80	Amusements	33	36	31	37	37
	75	Real Estate	33	34	33	34	34
	76	Hotels; Personal and Repair Services[c]	16	20	29	34	39
	78	Eating and Drinking Places	26	26	22	21	28
	84	Nonprofit Organizations	12	14	16	19	20
	72	Retail Trade	12	11	10	8	8
	83	Educational Services[c] (Private)	1	1	5	7	8
	82	Health Services[c] (except no. 81)	4	6	4	4	4
	81	Hospitals	0	0	0	0	0

NOTE: Horizontal rows divide the services roughly into three categories: (I) those delivered mainly to businesses; (II) those delivering about half their output to businesses, half to consumers; and (III) those delivering mainly to consumers.

[a] Services sectors are ordered by decreasing percentages in 1977.

[b] This column is not projected directly but is the result of computations involving assumptions about technological change.

[c] Indicates the six services whose use in production grew fastest between 1963 and 1977 (see Table 3).

SOURCE: Sector scheme and numbering correspond to those used in Leontief and Duchin (1986), which is also the source of the data base including the projections for 1990 (Institute for Economic Analysis data base). Sectors are defined on an industry basis. The original sources of the historical data are the U.S. Department of Commerce, Bureau of Economic Analysis and Bureau of Labor Statistics.

TABLE 2   Output and Use in Production of Selected Detailed Services in 1977 (expressed in 1977 prices)

Sector		Output ($ million)	Use in Production (%)
67[a]	Transportation; Warehousing	128,264	61
	Railroads and related services	22,462	74
	Local passenger transportation	10,899	31
	Motor freight transportation and warehousing	47,141	67
	Water transportation	16,873	53
	Air transportation	24,846	93
	Pipelines	3,346	91
	Freight forwarders	969	86
	Arrangement of passenger transportation	1,727	90
77[b]	Business Services	161,969	81
	Miscellaneous repair shops	10,262	86
	Services to buildings	5,260	80
	Personnel supply	5,035	68
	Computer and data processing	15,395	84
	Consulting and related	13,934	66
	Detective and protective	3,008	100
	Equipment leasing	9,850	92
	Photographic and related	3,744	51
	Other business services	9,037	93
	Advertising	36,292	97
	Legal	20,512	44
	Engineering and related	16,345	90
	Accounting and related	13,298	85

[a]Institute of Analysis sector 67 corresponds to Bureau of Economic Analysis (BEA) sectors 65.0100–65.0702.
[b]Institute of Analysis sector 77 corresponds to BEA sectors 73.0101–73.0303.
SOURCE: This table reports commodity output from the Bureau of Economic Analysis, U.S. Department of Commerce, 1984, Table 1. (Other tables in this paper describe industry output.)

Real growth rates of services sector outputs for 1963–1977 and projections for 1977–1990 are shown in Table 3 along with the growth of use in production alone. A comparison of Tables 3 and 1 shows that the services whose use in production grew fastest between 1963 and 1977 are those that deliver essentially to households: Private Education, Hotels and Personal and Repair Services, and Health Services. Payments for educational services by the private sector are projected to grow more slowly through 1990 because they are supplemented on a large scale by the in-house provision of services, accompanied by purchases of new products and services such as instructional television and computer-based instructional hardware or software (sectors 87 and 88 in Leontief and Duchin, 1986), which are not discussed here. The fast-growing business purchases of Hotels and Personal and Repair Services are essentially confined to hotels, laundry, and electrical repairs. The other

TABLE 3 Growth of Services Sector Output Between 1963 and 1977 and Projections for 1977 Through 1990 (average annual rate of real growth)

	Sector[a]	1963–1977 (%)		1977–1990 (%)		Output in 1977 (1979 $ million)[b]
		Overall	Use in Production	Overall	Use in Production	
68	Communications	8.2	7.3[c]	5.1	4.6[c]	54,733
81	Hospitals	7.2	None	2.5	None	92,346
77	Business Services	5.8	6.2[c]	3.9	3.6[c]	149,304
82	Health Services	5.8	5.7[c]	3.6	3.3	74,850
73	Finance	5.1	5.9[c]	3.8	3.3	85,968
75	Real Estate	5.0	5.1	3.2	3.3	338,020
79	Auto Repair Services	4.5	5.3	2.3	3.0	52,500
71	Wholesale Trade	4.4	4.9	2.7	2.8	188,780
74	Insurance	4.0	4.0	3.2	3.0	71,198
80	Amusements	3.9	4.7	3.7	3.7[c]	29,506
67	Transportation; Warehousing	3.8	3.6	2.8	2.7	145,548
70	Utilities	3.7	4.5	2.1	2.5	129,904
85	Government Postal Services	3.6	2.5	2.7	3.3	27,990
79	Eating and Drinking Places	3.4	1.8	2.0	4.4[c]	102,364
72	Retail Trade	3.3	0.7	2.7	2.8	195,038
69	Radio/TV Broadcasting	3.2	3.2	3.0	2.7	11,237
76	Hotels; Personal and Repair Services	3.1	8.5[c]	2.3	3.6[c]	61,017
84	Nonprofit Organizations	1.7	5.2	3.1	3.5[c]	29,501
83	Educational Services (Private)	−1.5	15.2[c]	1.5	2.4	12,195

[a]Sectors are ordered by descending overall growth rates for 1963–1977. Indicates the six services whose use in production grew fastest in each period.
[b]Growth rates are based on output valued in constant 1979 prices except for educational services that are valued in student-years. The deflators are discussed in Leontief and Duchin (1986). The last column is based on the industry outputs from the input-output table for 1977 estimated by the Bureau of Labor Statistics in 1981.
[c]Indicates the six services whose use in production grew fastest between 1963 and 1977.
SOURCE: Sector scheme and numbering correspond to those used in Leontief and Duchin (1986), which is also the source of the data base including the projections for 1990. Original sources of the historical data are the Bureau of Economic Analysis and the Bureau of Labor Statistics.

services whose use in production—and consumption—grew fastest are Communications, Business Services, and Finance.

THE REQUIREMENTS FOR MANUFACTURED GOODS
IN THE PRODUCTION OF SERVICES

A large assortment of manufactured inputs and raw materials are absorbed by the services sectors, some in only small quantities. Because of the highly developed division of labor, the links between these inputs and the services outputs are often indirect; for example, most of the processed metals required for the provision of Transportation Services are purchased and used in the sectors that produce the transportation equipment that is sold to establishments providing Transportation Services, whereas these establishments purchase little metal directly on their own account. In this chapter the total (i.e., direct plus indirect) input requirements for the delivery of services will be discussed.[3]

The economically most important total requirements to deliver $1,000 worth of Transportation Services, Communications Services, Insurance, Business Services, and Health Care in 1977 are shown in Table 4. Reading down the first column, for example, shows that significant amounts of Refined Petroleum ($127), Petroleum ($103),[4] Business Services ($67), and Motor Vehicles ($66) had to be produced (or imported) to deliver $1,000 of Transportation and Warehousing Services in 1977. The $1,160 of required Transportation and Warehousing services includes the $1,000 worth delivered to final users as well as $160 worth of output absorbed directly and indirectly, including in the production of the petroleum, motor vehicles, and so forth.

The Insurance sector stands out among the various services because it requires relatively small amounts of structures, energy, and transportation, and no special-purpose capital equipment. The major inputs are labor, paper, and assorted services. At the other extreme is the Transportation and Warehousing sector which, in addition to large energy requirements, requires a sufficiently large amount of transportation equipment for its indirect requirements of both ferrous and nonferrous metals to be significant: total requirements include $40 of steel versus only $10 of paper.

Table 5 has the same form as Table 4 and shows the corresponding growth rates between 1963 and 1977.[5] Because all production levels are evaluated in constant prices, these growth rates represent actual changes in the quantities of goods and services required and are not confounded by changes in their prices.

Of the ten largest requirements for the delivery of $1,000 worth of Insurance (from Table 4), only the major services inputs grew between 1963 and 1977; all the others declined. In the case of Communications Services, falling requirements for Structures, Energy, Transportation, Paper, Metals, and even for most services were partly offset by increased use of Communications Equipment and Business Services.

Transportation Services, Business Services, and Health Care all show significant increases in virtually all of the important requirements except labor. In the latter two cases it is reasonable to interpret these increases as a change in the nature of the services that is not reflected in the price deflators. Some of the greatest increases are experienced in services inputs: Communications, Wholesale Trade, Finance, Business Services, and Insurance.

We can conclude that the services sectors make intensive use of buildings, energy, and paper; they also rely on the materials-processing sectors for transportation equipment and special-purpose machinery. Aside from refined petroleum and papermaking, dependence on the chemical sectors is mainly for pharmaceuticals. Examination of the largest requirements does not reveal a systematic increase or decrease in the changing use of manufactured inputs, although it will be important to follow up this analysis with an investigation of the implications of ongoing computerization for the use of materials inputs.

THE IMPORTANCE OF SERVICES INPUTS FOR MANUFACTURING

The importance of purchased services for the delivery of manufactured goods can now be examined. Four manufacturing and four services sectors, representative of the range of production characteristics of manufacturing and services sectors, are compared in Table 6 with respect to their use of selected services in 1977. Transportation and Warehousing Services, Utilities, and Wholesale Trade were used more intensively in the delivery of all manufactured goods than in the delivery of any of the services, whereas Communications Services, Finance, Insurance, and Government Postal Services appear to be equally important in the production of goods and services.

Changes between 1963 and 1977 in the requirements for these services are shown in Table 7. The most striking findings in this table are the nearly sixfold increase in the use of insurance coverage to provide a fixed quantity (i.e., that which could be purchased in 1977 for $1,000 in the prices of 1979) of health care services and the reduction, by similar orders of magnitude, in requirements for all of the services shown in the table in order to deliver a given quantity of computer capability. The former reflects mainly social changes, while the latter results from dramatic advances in production technology and in the scale of production of electronic components and computers. It turns out that a similar magnitude of increase occurred in the purchase of Business Services for the provision of Agricultural Services.

According to Table 7, the use of Insurance in the other sectors has not changed much per unit of delivered output, and the use of Government Postal Services declined noticeably in some, and increased in other, sectors. By and large, however, services inputs have increased.

TABLE 4 Production Required to Deliver $1,000 of Output of Five Selected Services to Final Users in 1977 (all quantities measured in constant 1979 prices)

	67 Transportation Services	68 Communications	74 Insurance	77 Business Services	82 Health Care (except Hospitals)
Structures					
11 Construction	92	92	22	26	30
75 Real Estate	47	41	56	67	58
Energy					
8 Petroleum	103	(14)	(16)	25	(17)
30 Petroleum Refining	127	(16)	(20)	32	19
70 Utilities	(28)	21	(19)	(19)	23
Transportation					
61 Motor Vehicles	66	(10)	(11)	20	*
62 Aircraft	39	*	*	*	*
63 Other Transportation Equipment	44	*	*	*	*
67 Transportation Services	1,160	20	22	38	(18)
79 Auto Repair	(30)	(8)	(13)	(19)	(8)
Other Services					
68 Communications Services	(18)	1,052	37	24	(10)
71 Wholesale Trade	51	21	20	28	29
73 Finance	(26)	18	59	(18)	(10)
74 Insurance	(25)	(7)	1,398	(10)	23
77 Business Services	67	47	82	1,099	54

TABLE 4 Continued

Paper					
23 Paper	(10)	(10)	20	(13)	(10)
25 Printing and Publishing	(14)	22	32	25	(13)
85 Government Postal Services	(13)	(12)	(19)	(16)	(13)
Other					
36 Steel	40	(15)	*	(15)	*
37 Nonferrous Metals	(23)	21	*	(11)	*
56 Communications Equipment	*	93	*	*	*
64 Scientific Instruments	*	*	(18)	*	(16)
13 Food	(11)	*	*	*	22
28 Drugs, etc.	*	*	*	*	21
76 Hotels; Personal and Repair Services	*	(13)	(13)	20	20
78 Eating Places	(17)	(12)	40	(19)	(14)
82 Health Care (except Hospitals)	*	*	*	*	1,036

NOTE: The ten largest production requirements in 1977, in addition to the delivering sector's own output, are shown in each column. Quantities of less than $10 are indicated by an asterisk; those greater than or equal to $10, but not among the ten largest, are enclosed in parentheses. Also categories shown in all cases include the ten largest production requirements in 1963. Figures correspond to input-output "total requirements" including direct and indirect requirements for current-account inputs and replacement capital.

SOURCE: Institute for Economic Analysis data base.

TABLE 5 Changes in Production Requirements to Deliver $1,000 of Output of Five Selected Services to Final Users Between 1963 and 1977 (1977 requirement divided by 1963 requirement, both in constant prices)

	67 Transportation Services	68 Communications	74 Insurance	77 Business Services	82 Health Care (except Hospitals)
Structures					
11 Construction	.92	.61	.54	1.08	1.11
75 Real Estate	1.18	.91	.72	1.31	1.41
Energy					
8 Petroleum	1.69	.54	1.07	1.79	1.89
30 Petroleum Refining	1.23	.41	1.11	1.78	1.73
70 Utilities	1.33	.57	.49	.68	1.00
Transportation					
61 Motor Vehicles	2.13	.56	2.75	3.33	*
62 Aircraft	2.44	*	*	*	*
63 Other Transportation Equipment	1.05	*	*	*	*
67 Transportation Services	1.03	.67	.85	1.27	1.20
79 Auto Repair	1.25	.40	1.86	2.11	2.67
Other Services					
68 Communications Services	2.00	1.03	1.61	2.67	1.67
71 Wholesale Trade	1.34	.68	1.11	1.65	1.93
73 Finance	2.00	.95	1.48	2.25	2.00
74 Insurance	1.09	.58	1.05	1.43	5.75
77 Business Services	1.86	1.12	.93	1.03	3.00

TABLE 5 Continued

Paper					
23 Paper	1.25	.77	.63	1.00	1.43
25 Printing and Publishing	1.17	1.00	.64	1.14	1.63
85 Government Postal Services	.93	.50	.61	.37	.93
Other					
36 Steel	.87	.55	.73	1.25	*
37 Nonferrous Metals	1.10	.75	.54	1.22	*
56 Communications Equipment	*	1.24	*	*	*
64 Scientific Instruments	*	*	*	*	1.33
13 Food	1.00	*	.82	.64	1.47
28 Drugs	*	*	*	*	1.50
76 Hotels; Personal and Repair Services	*	2.60	2.17	2.86	2.22
78 Eating Places	1.13	.67	.93	.90	.90
82 Health Care (except Hospitals)	*	*	*	*	1.02

NOTE: Asterisk indicates that less than $10 was required in both 1963 and 1977 to deliver $1,000 of the service to final users (in 1979 prices).
SOURCE: Table 4 and Institute for Economic Analysis data base.

TABLE 6 Total Requirements of Seven Services to Deliver $1,000 of Selected Manufactured Goods and of Selected Services to Final Users in 1977 (1979 prices)

Delivered to	Services from						
	67 Transportation, Warehousing	68 Communications	70 Utilities	71 Wholesale Trade	73 Finance	74 Insurance	85 Government Postal Services
17 Apparel	60	13	53	96	21	9	16
26 Chemicals	95	11	120	62	20	10	20
36 Steel	105	10	102	91	17	11	18
50 Computers	80	23	46	96	39	13	17
73 Finance	19	27	28	17	1,132	10	41
74 Insurance	22	37	19	5	59	1,398	19
76 Hotels; Personal and Repair Services	39	20	46	54	22	16	15
82 Health Services (except Hospitals)	18	10	23	29	10	23	13

NOTE: Some entries in this table and in Tables 8 and 10 are greater than $1,000. For example, to deliver $1,000 worth of Financial Services to final users, some intraindustry purchases are made within the producing sector. The use of financial services by the sectors producing the structures, energy, etc., for Finance (indirect requirements) is also counted. This direct and indirect usage adds $132 to the $1,000 of deliveries.

SOURCE: Institute for Economic Analysis data base.

TABLE 7 Changes in Requirements of Seven Services to Deliver $1,000 of Selected Manufactured Goods and Selected Services to Final Users Between 1963 and 1977 (1977 requirement divided by 1963 requirement, both in 1979 prices)

Delivered to	Services from						
	67 Transportation, Warehousing	68 Communications	70 Utilities	71 Wholesale Trade	73 Finance	74 Insurance	85 Government Postal Services
17 Apparel	1.07	1.44	1.51	1.02	1.31	.75	.89
26 Chemicals	1.53	1.83	1.46	1.32	1.82	1.00	1.33
36 Steel	1.48	2.00	1.89	2.17	1.55	1.00	1.64
50 Computers	.23	.24	.22	.26	.27	.15	.15
73 Finance	1.00	1.69	.97	1.42	1.03	1.11	.91
74 Insurance	.85	1.61	.49	.28	1.48	1.05	.61
76 Hotels; Personal and Repair Services							
82 Health Services (except Hospitals)	1.23	2.86	1.77	1.64	1.83	1.14	1.15
	1.20	1.67	1.00	1.93	2.00	5.75	.93

SOURCE: Table 6 and Institute for Economic Analysis data base.

Significant findings about the most important users of eight of the services sectors are highlighted in the next two tables. One finding is that these services can be divided into two groups: the first (shown in Table 8) is characterized in part by a significantly larger dollar volume of output (see Table 3) and a significantly larger average use by other sectors per dollar of their deliveries (compare Tables 8 and 9) than the second.

Deliveries of nondurable manufactured goods and processed materials account for the largest interindustry requirements of Transportation Services, Utilities, and Wholesale Trade. For example, Table 8 shows that delivery of $1,000 worth of Paper and Paper Products in 1977 required $103 of Transportation and Warehousing Services and $94 of Wholesale Trade (as well as $85 of Utilities that is not shown), while delivery of $1,000 of Stone and Clay products required $143 of Transportation and Warehousing Services and $89 of Utilities (and $59 of Wholesale Trade, not shown) (all in 1979 prices). Although the typical requirements for Business Services are of the same dollar volume, the principal users per dollar of deliveries are the electronics sectors: Computers, Office Equipment, Communications Equipment, Semiconductors, and Other Electronic Devices. It is notable that the services sectors are not among the principal users of the services shown in this table.

Communications Services, Finance, Insurance, and Government Postal Services are required in relatively small dollar amounts for the delivery of most outputs—a high of $20–$30 worth per $1,000 of deliveries, contrasted with about $100 worth for the last group of services discussed. Table 9 shows that six of the ten most important users of Communications Services per dollar of deliveries in 1977 were services: Insurance, Radio and TV Broadcasting, Finance, Wholesale Trade, Business Services, and Nonprofit Organizations, in addition to Communications Services itself ($52 worth of Communications Services were absorbed in the process of producing $1,000 worth for delivery to final users). Services sectors were also six of the ten most important users of Insurance, while Finance and Government Postal Services were used most intensively in the delivery of goods. The use of all the services shown in this table by many other sectors (not shown) is almost as great as their use among the top ten; this contrasts with the greater concentration of intensive use of Transportation and Warehousing, Utilities, and Wholesale Trade.

The use of Insurance and of Government Postal Services per unit of deliveries fell off noticeably from 1963 to 1977 for a number of sectors, while the use of Communications and Finance by most sectors increased. There was no doubt some direct substitution of other forms of communication for the U.S. mail.

TABLE 8 Principal Uses of Transportation and Warehousing Services, Utilities, Wholesale Trade, and Business Services in 1977 (total requirements of the four services to deliver $1,000 of output of each named sector to final users, in 1979 prices)

67 Transportation Services		70 Utilities		71 Wholesale Trade		77 Business Services	
67 Transportation Services	1160	70 Utilities	1363	71 Wholesale Trade	1035	77 Business Services	1099
35 Stone and Clay Products	143	10 Chemical Mining	178	16 Textiles	115	73 Finance	138
24 Paper Boxes	111	5 Iron Mining	134	61 Motor Vehicles	107	28 Drugs, etc.	125
29 Paints	107	85 Government Postal Services	124	20 Wood Containers	105	58 Semiconductors	108
36 Steel	105	26 Chemicals	120	15 Fabrics	103	4 Agricultural Services	105
23 Paper	103	27 Plastics	110	78 Eating Places	102	51 Office Equipment	99
27 Plastics	102	34 Glass	105	18 Miscellaneous Textiles	102	59 Other Electronics	98
38 Metal Containers	98	36 Steel	102	17 Apparel	96	56 Communications Equipment	95
16 Textiles	96	37 Nonferrous Metals	99	50 Computers	96	69 Radio/TV	93
20 Wood Containers	95	9 Stone Quarrying	92	21 Household Furnishings	95	50 Computers	87
26 Chemicals	95	35 Stone and Clay Products	89	23 Paper	95	84 Nonprofit Organizations	85
50 Computers	80	58 Semiconductors	68	54 Appliances	94	74 Insurance	82
13 Food Processing	80	57 Electron Tubes	65	52 Services Industry Machinery	89	57 Electron Tubes	74
59 Other Electronics	75	50 Computers	46	59 Other Electronics	68	54 Appliances	65
85 Government Postal Services	74			58 Semiconductors	65	71 Wholesale Trade	61
58 Semiconductors	71			57 Electron Tubes	46		
54 Appliances	68						

NOTE: The first ten sectors named in each column are the largest users in 1977 (after the column sector itself). The subsequent sectors in each column are the remainder of the ten largest users in 1963 whose relative importance diminished by 1977.

SOURCE: Institute for Economic Analysis data base.

TABLE 9 Principal Uses of Communications Services, Finance, Insurance, and Government Postal Services in 1977 (total requirements to deliver $1,000 of output of each named sector to final users, in 1979 prices)

68 Communications		73 Finance		74 Insurance		85 Government Postal Services	
68 Communications	1,052	73 Finance	1,398	74 Insurance	1,132	85 Government Postal Services	1,020
74 Insurance	37	74 Insurance	59	67 Transportation Services	59	73 Finance	41
69 Radio/TV	31	50 Computers	39	82 Health Care	39	84 Nonprofit Organizations	33
73 Finance	27	20 Wood Containers	37	75 Real Estate	37	70 Utilities	27
71 Wholesale Trade	24	58 Semiconductors	31	80 Amusements	31	25 Printing, Publishing	26
77 Business Services	24	59 Other Electronics	31	79 Auto Repair	31	10 Chemical Mining	25
50 Computers	23	65 Optical Equipment	30	2 Agricultural Products	30	5 Iron Mining	25
84 Nonprofit Organizations	21	57 Electron Tubes	29	1 Livestock	29	27 Plastics	25
58 Semiconductors	19	60 Electrical Supplies	27	4 Agricultural Services	27	16 Textile Goods	21
59 Other Electronics	19	66 Miscellaneous Manufactures	27	76 Hotels; Personal and Repair Services	27	26 Chemicals	20
28 Drugs, etc.	19	10 Chemical Mining	27	9 Stone Quarrying	27	23 Paper	20
64 Scientific Instruments	16	9 Stone Quarrying	22	13 Food	22	74 Insurance	19
57 Electron Tubes	14	6 Nonferrous Mining	21	21 Household Furnishings	21	58 Semiconductors	19
54 Appliances	13	16 Textiles	21	50 Computers	21	59 Other Electronics	19
		28 Drugs, etc.	20	71 Wholesale Trade	21	50 Computers	19
		68 Communications	18	58 Semiconductors	20	77 Business Services	18
				59 Other Electronics	18	57 Electron Tubes	17
						68 Communications	16

NOTE: The first ten sectors named in each column are the largest users in 1977 (after the column sector itself). The subsequent sectors in each column are the remainder of the ten largest users in 1963 whose relative importance diminished in 1977.

SOURCE: Institute for Economic Analysis data base.

THE CAPTIVE PRODUCTION OF SERVICES BY BUSINESSES

The Business Services sector produced about $135 billion of output for sale in 1977 and employed over four million people. Like many of the services sectors, this sector [as defined on a two-digit Standard Industrial Classification (SIC) basis] is far larger than the average two-digit manufacturing sector: a single component, Advertising, had as large a volume of sales in 1977 as the entire paper industry. An adequate analysis of the changing role of services will require a far more detailed representation of the services activities in the official (and other) data series. Disaggregation will also contribute to improving the price deflators for services sectors: since the output of a more narrowly defined sector is more homogeneous, one can more readily define a unit of output (and consequently a unit price) that corresponds to actual transactions (e.g., visits to a dentist's office rather than health care services).

Although this type of attention to the services sectors is overdue and will be forthcoming, a more fundamental issue needs to be addressed regarding the preparation of data for analyzing the role of the services sectors. The $135 billion of Business Services and four million employees are the tip of the iceberg: a large proportion of U.S. workers are engaged in the production of services for business, notably most managers and clerical workers. If they produce personnel, legal, data processing, or other services for sale by a specialized establishment, this work is counted as services output. If they perform a similar function within, IBM or General Motors, the official (and other) data series do not record this work as the production of services. In the latter case these services represent secondary, and furthermore captive, production.

It is useful to examine the treatment of secondary and captive production in official data series in the case of physical outputs because the distinctions are more familiar in this context. Secondary production records outputs—other than the primary product—that are produced for sale or, in some cases, for transfer to another establishment within the same firm. Several "establishments" may be defined at the same geographic location, one corresponding to each "product" produced; the outcome of some stage of production may be considered a product if it is the primary output of some other sector. One establishment is then said to transfer some of its output to another at the same address, as a device for recording captive production. Thus, for example, construction performed within a manufacturing establishment is "redefined" as output purchased from the construction industry, and steel produced within automobile plants is added to that produced within the steel industry. In these cases, not only the inputs but also the outputs of captive production (i.e., construction activity and steel production) are explicitly accounted for.

Even in the case of physical production, not all such activity is captured for a variety of reasons. For example, the cogeneration of electricity and the captive production of robots are not recorded (there is not even an SIC code for robots) because they have only relatively recently attained economic importance and because there is no automatic mechanism for deciding when to identify the outputs of various stages of production as secondary products. However, the captive production of services is not recorded at all.

A large volume of the intrafirm transfer of services is from corporate headquarters (and other "administrative" units such as R&D facilities) to operating establishments. In industry data series, the inputs of these administrative units are allocated among the establishments they serve. This is represented as an increase in the operating unit's purchases of personnel, structures, paper, and so on—not as a transfer of accounting services or data processing services. Other types of services are performed within virtually every establishment. Because the production for an establishment's own use of services has not been recorded in the past, and is not recorded now, it is not possible to conclude to what extent the increased purchases and sales of services (that were quantified in the last section) reflect an increased inter-industry division of labor based on the specialized production of services or to what extent they represent the increased use of services.

There are many unresolved issues surrounding the systematic and comprehensive recording of secondary, especially captive, production. The discussion here is limited to identifying some guidelines for the classification of services that will make possible a more even-handed treatment of the captive production of physical goods and of services.

Physical goods produced for an establishment's own use bear a strong resemblance to those produced for sale, as a primary or a secondary product, both in the properties of the finished product and, often, in the process of production. The decision to produce for an establishment's own use is generally the outcome of an explicit comparison of the relative costs and other attributes of "make versus buy." Management is at a much earlier stage in the systematic comparison of making versus buying those services required for production that have typically been produced in-house. The nature of such a service and its production process are likely to be different from those of specialized services establishments. However, the preoccupation with cost reduction and the rationalization associated with automation are powerful pressures for the systematic assessment of these trade-offs and for the standardization of many types of services and their production. At the same time, these same pressures lie behind the growing importance of specialization in the provision of business services.

The business processes (nonmanufacturing) generally performed within a firm are listed in Table 10. It is instructive to compare this classification scheme with the 13 subcategories of Business Services (SIC-based) shown

TABLE 10 Process Classification Scheme for Corporate Services and Auxiliary Support

1. Finance	8. Health/Education
1.1 Accounting	8.1 Counseling
1.2 Financial Analysis	8.2 Education
2. Administration	8.3 Health Care
2.1 Legal	9. Operations Support
2.2 Personnel	9.1 Inspection
2.3 Public Relations	9.2 Manufacturing Support
3. Planning	9.3 Purchasing
3.1 Planning, Marketing, Business Analysis	9.4 Quality Control
3.2 General Management	9.5 Shipping, Receiving
3.3 Project Management, Project Office	9.6 Warehousing
4. Office Production	10. Services
4.1 Publications	10.1 Entertainment
4.2 Office Support, Secretarial	10.2 Food Services
4.3 Customer Relations	10.3 Postal Services, Mailroom
4.4 Application Evaluation	10.4 Personal Services
5. Professional Sales	10.5 Security, Safety
5.1 Professional Sales	11. Maintenance
6. Technical	11.1 General Maintenance
6.1 Laboratory Analysis	11.2 Office Equipment Maintenance
6.2 Research and Development	12. Other
6.3 Telecommunications	12.1 Construction
7. Computer Operations	12.2 Transportation of Persons
7.1 Computer Operations	12.3 Transportation of Product
7.2 Data Entry	
7.3 Systems Analysis, Design, Programming	

SOURCE: Duchin (1988a, p. 104). This classification draws on the Work Group Analysis done at IBM.

in Table 2 and the entire range of services industries shown in Table 1; their reconciliation remains to be resolved. The range of services carried out within firms will eventually need to be reflected in the services distinguished in the SIC. Once a unified set of services has been defined, the decision to make or buy these services can be analyzed in terms of the technical and business feasibility of separating their execution from other internal activities and the degree of standardization that can be achieved to assure their integration with these other activities, as well as in terms of relative costs. Basing the new services classification not only on accounting but also on analytical considerations such as these will improve the usefulness of the official data series both as a record of the past and as input to current decisionmaking.

Captive production is a characteristic, not an incidental, attribute of business services, yet it has gone completely unrecorded until now. Identification of the individual activities carried out in production provides the framework necessary for recording how these activities are accomplished through some combination of captive production and purchase of goods and services. These processes play another analytical role that was, in fact, the initial motivation for identifying them: to provide a framework for the systematic projection of alternative future input-output structures based on the conceptually most plausible sources of information. For this discussion the reader is referred to Duchin (1988a).

HOUSEHOLD PRODUCTION AND PURCHASES OF SERVICES

In basic outline, the discussion of make or buy decisions for businesses also applies to households. However, the representation in official data series of household use and purchases of goods—not to mention services—is far more rudimentary than that of businesses; and recommendations for redressing this imbalance have been described elsewhere (Duchin, 1988b). The first need is for a household classification scheme (eventually a standard household classification, SHC), whose nonexistence has until now caused surprisingly little consternation on the part of analysts. A second requirement is a classification of household activities similar to that shown for business activities in Table 10. A preliminary classification scheme is proposed in Table 11: different categories of households can be expected to execute these processes through characteristic combinations of purchasing versus their own production of various goods and services. The captive production of business services has a close analogue in unpaid household labor: what households do for themselves must be examined as a basis for understanding their changing patterns of purchases.

The detailed classification and description of households and their activities will make it possible to analyze the changing use of services by different categories of households, just as their changing use by different categories

---

TABLE 11   Proposed Classification Scheme for Activities of Households

1. Food (obtain, store, prepare, cleanup)
2. Clothing (obtain, maintain, clean)
3. Health Care (visit provider, obtain products)
4. Living Quarters (obtain, maintain, clean)
5. Child Care
6. Personal Grooming
7. Education
8. Recreation
9. Transportation
10. Household Management (recordkeeping, tax preparation, etc.)
11. Activities Related to Paid Employment

of industrial sectors is analyzed in previous sections of this chapter. In both cases, an additional level of structure in terms of activities, such as those suggested by Tables 10 and 11, will contribute to the explanation of observed changes. The empirical analysis of changing production and household activities can then quantify and anticipate the shift (when it occurs) from goods and services produced for oneself (captive production) to those that are purchased.

IMPLICATIONS OF THE ANALYSIS

A detailed examination of the services sectors reveals their great diversity and their characteristic interdependence with manufacturing and other materials-processing sectors. The growth of the services sectors has been accompanied by significant demand for construction, energy, paper, transportation equipment, and various special-purpose capital goods that are among the largest inputs (in value) to virtually all the services sectors. The notion that services involve essentially people (and computers), and not manufactured inputs, turns out to be an unrealistic basis for policy.

While the common perception is valid that many services provide relatively small deliveries to all sectors, the markets for several of the largest services sectors consist almost exclusively of manufacturing establishments. There is surely competition between services and manufacturing sectors (in fact, among all sectors) for investment dollars and preferential treatment of various sorts; however, the weakening of the U.S. manufacturing base probably harms services sectors more than it helps them because the manufacturing sectors not only provide inputs but also absorb outputs of services.

An initial assessment of the interconnections of household and services sector activities identified several large services sectors that divide their output about equally between households and businesses. It is important not to obscure this overlap by the separate treatment of producers' services and consumers' services, and to recognize technological and organizational changes taking place in the home as well as in the office and factory.

Talk of the "postindustrial era" and "age of information" notwithstanding, our society still sees fit to invest enormous sums, as a matter of national strategy and policy, in research in the natural sciences that is intended to pay off in the form of physical, manufactured products such as new materials and vaccines. Comparably important results affecting our economic and social well-being still await (1) acknowledging the potential importance of the "service" provided by an economic analysis and (2) allocating the order of magnitude of funding that could make it feasible to build and use modern tools (including data bases and models) for a suitably detailed description and analysis of the past, as well as an evaluation of our prospects and alternatives for the future.

NOTES

1. Throughout this chaper, investment goods are considered as an input to production, rather than being conventionally included with consumption as "final demand.")
2. Net trade is included in the latter, accounting for a very small proportion of the total in all cases. It is not possible to describe the numerous definitions and conventions that underlie the figures in this and the following tables. For example, certain purchases of wholesale trade services are charged directly to households, resulting in what may appear to be a surprisingly small proportion of sales to business. Nor can the various discrepancies among numbers from different sources and those reported in these tables be discussed. This chaper makes extensive use of the Institute for Economic Analysis (IEA) input-output data base. Despite its deficiencies (e.g., all values are in 1979 rather than more recent prices, and the most up-to-date official data have not yet been incorporated), it is still the only data base on which the analysis reported in this paper can be based. In addition to compatible tables for all benchmark years, the data base includes detailed matrices of labor requirements and of capital requirements for the replacement and the expansion of production capacity.
3. The total requirements matrix used here is computed as $(I - A - R)^{-1}$, where I is the identity matrix, A is the current account input-output matrix, and the matrix R describes replacement of capital. The tables in sections 3 and 4 contain portions of the rows and columns, respectively, of this inverse matrix.
4. The value of refined Petroleum ($127) includes the value of the corresponding crude; the latter is also included in Petroleum requirements ($103). These two—or any other—total requirements should of course not be summed because of an indeterminate amount of such "double-counting."
5. In most but not all cases, there is a monotonic or nearly monotonic movement in these total requirements from 1963 to 1967 to 1972 to 1977, the four benchmark years for official input-output tables.

REFERENCES

Duchin, F. 1988a. Analyzing technological change: An engineering data base for input-output models of the economy. Engineering with Computers 4:99–105.

Duchin, F. 1988b. Analyzing structural change in the economy in input-output analysis: Current developments, Ciaschini, ed. London: Chapman and Hall.

Fourastie, J. 1966. Idées Majeures. Paris: Éditions Gonthier. (translation from p. 32).

Leontief, W., and F. Duchin. 1986. The Future Impact of Automation on Workers. New York: Oxford University Press.

U.S. Department of Commerce, Bureau of Economic Analysis. 1984. The Detailed Input-Output Structure of the U.S. Economy, 1977, Vol. 1. Washington, D.C.

Productivity in Services

JOHN W. KENDRICK

The chief factor behind secular increases in productivity—the ratio of output to inputs in real terms—is innovation in the technology and organization of production. Other factors, such as cyclical changes in the rate of utilization of capacity, are also important, as discussed later. However, in the long run, it is technological advance that is primarily responsible for reducing real costs per unit of output (the opposite side of the productivity coin). This is obviously so in the case of process innovations or new and improved producers' goods. Even the introduction and diffusion of new consumer goods have a positive influence on productivity through the learning curve effect.

In the first section of this chapter, estimates of rates of change in productivity in the goods and services sectors of the U.S. business economy by major industry groups, 1948–1986, are presented. This furnishes the basis for analysis of causal factors behind productivity in the second section by using two of the three major approaches available. The first is growth accounting by which the chief forces promoting productivity growth in the total economy are identified and evaluated. The second is the use of multiple regression analysis to relate relative changes in productivity by industry to differences in levels or rates of change in causal variables.

There is a third approach at the microlevel which uses case studies of individual firms and technologies. This approach is well represented in the companion book to this volume (Guile and Quinn, 1988). Additionally, the American Productivity Center in Houston has published more than 50 case studies of productivity in various companies covering innovation, human resources management, and other factors.[1] There have also been interfirm and interplant comparisons of productivity levels and rates of change that have generally confirmed the findings of macroeconomic analyses.

The third and final section of this chapter presents a menu of policy options for promoting productivity growth. These are generally policies that would stimulate productivity across the board, but some are of particular relevance to the services sector as pointed out in the concluding comments.

OUTPUT, INPUT, AND PRODUCTIVITY DEVELOPMENTS, 1948–1986

Sector Comparisons

The marked increase in the share of the services sector (defined in terms of the criterion of tangibility), both in persons engaged and in real gross product between 1948 and 1986, is shown in Table 1. Persons engaged include proprietors and full-time equivalent employees. Real gross product is the value added by the various industry groups in constant 1982 dollars, as estimated by the Bureau of Economic Analysis in the U.S. Department of Commerce.

The proportion of all persons engaged in the U.S. economy working in services, private and public, rose from 54 percent in 1948 to more than 72 percent in 1986. This continues a trend that has been evident since at least 1870, when 23 percent of the labor force was engaged in services. The share of real gross product originating in services has also increased, but to a somewhat lesser degree, as shown in Table 1. This implies that real product per person engaged rose less rapidly in services than in goods production.

An alternative picture of sectorial shares is provided by the breakdown of final goods and services entering the gross national product (GNP), shown in Table 2. The proportion of tangible goods declined from 48 percent in 1948 to approximately 43 percent in 1973 and 1986. The share of final services grew from 39.5 percent in 1948 to 46.6 percent in 1986. The share of structures remained close to 12.5 percent in 1948 and 1973, declining to 10.4 percent in 1986. This sectoring gives somewhat different results from the industry sectoring because it is based on final products, and some of the services industries add value to tangible goods. However, Table 2 confirms the finding that services constitute an increasing proportion of GNP.

Returning to the broad sectorial aggregations of industries, Table 3(A) shows that real gross product in services grew more than 1 percent faster per year, on the average, than in goods production between 1948 and 1986— 3.7 compared with 2.6 percent. The slowdown of growth after 1973 was less pronounced in services than in goods.

Table 3(B) shows the relative performance of total factor productivity (TFP). As a measure of the net savings in labor plus capital inputs per unit

TABLE 1   Real Gross Domestic Product and Persons Engaged in U.S. Domestic Economy by Sector and Industry Group

	1948	1973	1986
A. Persons engaged			
Total (thousands)	58,301	83,299	104,645
Percentage distribution			
Agriculture	11.4	3.8	2.8
Mining,	1.8	0.8	0.8
Construction	5.7	5.8	5.8
Manufacturing	27.4	23.9	18.0
Total Goods	**46.2**	**34.2**	**27.4**
Transportation	5.2	3.4	3.1
Communications	1.3	1.3	1.1
Public utilities	0.9	0.9	0.9
Trade	18.3	19.5	21.2
Finance, insurance, real estate	3.2	5.0	6.3
Services	13.2	17.9	23.7
Total Private Services	**42.1**	**48.0**	**56.4**
Government	**11.7**	**17.8**	**16.2**
B. Real gross domestic product			
Billions of 1982 dollars	1,114.9	2,726.6	3,694.5
Pecentage distribution			
Agriculture	5.5	2.6	2.7
Mining	6.5	4.9	3.2
Construction	8.1	6.2	4.6
Manufacturing	21.4	22.8	22.0
Total Goods	**41.5**	**36.5**	**32.5**
Transportation	6.9	4.3	3.5
Communications	0.8	1.9	2.6
Public utilities	1.2	2.7	2.8
Trade	14.5	16.0	17.4
Finance, insurance, real estate	9.7	13.5	13.9
Services	11.6	12.5	15.3
Total Private Services	**44.6**	**50.9**	**56.5**
Government	**13.9**	**12.6**	**11.0**

SOURCE: Bureau of Economic Analysis, U.S. Department of Commerce.

of output over time, TFP indicates changes in productive efficiency. Over the long term, increases in TFP result primarily from innovations in the technology of production, which are usually incorporated in capital goods. Organizational changes, including those associated with increasing scale, are also important. Between 1948 and 1986, the average annual percentage rate of increase in TFP in services was 1.4 percent, compared with 2.1 percent in goods production. The slowdown in goods productivity from 1973 to 1979

TABLE 2   Distribution of Real Gross National Product by Major
Category of Final Product

	1948	1973	1986
Billions of 1982 Dollars			
Total GNP	1,109	2,744	3,713
Goods	532	1,175	1,595
Services	438	1,219	1,731
Structure	139	350	387
Percentage Distribution			
Total GNP	100.0	100.0	100.0
Goods	48.0	42.8	43.0
Services	39.5	44.4	46.6
Structures	12.5	12.8	10.4

SOURCE: Bureau of Economic Analysis, U.S. Department of Commerce.

was greater than in services. However, services productivity growth fell a bit further from 1979 to 1986, whereas goods productivity accelerated smartly due mainly to comebacks in manufacturing and farming.

The reconciliation item between rates of change in TFP and in the more conventional real product per labor hour ("labor productivity") is the rate of substitution of capital for labor. This is indicated by the rates of change in the capital/labor ratios shown in Table 3(C) when they are weighted (multiplied) by the capital share of gross product, which averages about one-third. The capital/labor ratio increased much more in goods than in services between 1948 and 1986—2.3 percent a year, on average, compared with 1.3 percent. There was a slowing in the 1973–1979 subperiod but a comeback after 1979 in both sectors. The slower growth of capital (primarily structures and equipment) in services has important policy implications, which are discussed later.

Labor productivity increased at a 2.2 percent average annual rate in the U.S. business economy between 1948 and 1986, compared with 1.7 percent for TFP, the difference reflecting the growth of capital goods per worker. The increase in goods productivity averaged 2.7 percent annually, and in services 1.9 percent. As in the case of TFP, there was a sharp deceleration in labor productivity from 1973 to 1979, with a comeback in the goods sector from 1979 to 1986, in contrast to continued slow growth in the services sector as a whole.

An important caveat must be attached to the output and productivity estimates based on real gross domestic product (GDP) by industry as estimated by the Department of Commerce. As documented elsewhere (Kendrick, 1986a), about half of the base-period value added in the finance and private services industry groups has, in effect, been extrapolated by employment or

TABLE 3 Output and Productivity Ratios in U.S. Business Economy by Major Sector and Industry Group (average annual percentage rates of change, 1948–1986, by subperiod)

	1948–1985	1948–1973	1973–1979	1979–1986
A. Real Gross Product in 1982 Dollars				
Business economy	3.2	3.6	2.5	2.5
Goods production	2.6	3.2	1.3	1.6
Farming	1.4	0.4	1.5	5.0
Manufacturing	3.3	3.9	1.9	2.2
Mining	1.3	2.5	−0.4	−1.4
Construction	1.7	2.6	0.6	−0.4
Services production	3.7	4.0	3.4	3.1
Transportation	1.4	1.7	2.7	−0.9
Communications	6.4	7.1	6.2	4.0
Public utilities	5.6	7.2	1.8	3.2
Trade	3.7	4.1	2.7	3.4
Finance and insurance	4.1	4.3	3.4	3.9
Real estate	3.7	4.2	4.3	1.4
Services	4.1	4.0	4.4	4.3
B. Total Factor Productivity				
Business economy	1.7	2.3	0.3	0.8
Goods production	2.1	2.7	0.1	1.8
Farming	3.6	3.2	1.8	6.7
Manufacturing	2.1	2.3	0.6	2.5
Mining	0.7	2.7	−6.0	−0.5
Construction	−0.3	0.6	−2.2	−1.9
Services production	1.4	2.0	0.5	−0.3
Transportation	1.6	2.3	1.5	−0.5
Communications	3.8	4.9	2.4	1.3
Public utilities	3.3	4.9	−0.6	0.9
Trade	2.0	2.5	0.4	1.4
Finance and insurance	0.4	1.1	−0.7	−1.0
Real estate	0.7	1.3	1.4	−2.0
Services	0.6	0.7	0.4	0.3
C. Capital/Labor Ratios				
Business economy	1.9	2.0	1.2	1.7
Goods production	2.3	2.4	1.9	2.2
Farming	4.2	5.2	2.6	2.2
Manufacturing	2.8	2.6	3.2	3.2
Mining	4.1	5.1	−3.3	7.1
Construction	1.3	2.3	1.5	−2.3
Services production	1.3	1.5	0.6	1.2
Transportation	0.2	0.2	−0.1	0.3
Communications	5.0	5.2	4.0	5.1
Public utilities	2.5	3.3	1.5	0.4
Trade	3.0	3.1	2.4	3.2
Finance and insurance	2.9	1.9	4.0	6.3
Real estate	0.5	0.7	−1.7	1.5
Services	1.0	1.8	0.5	−0.7

continued

TABLE 3   Continued

	1948–1985	1948–1973	1973–1979	1979–1986
D. Real Product per Labor Hour				
Business economy	2.2	2.8	0.6	1.4
Goods production	2.7	3.2	0.5	2.5
Farming	5.0	4.7	3.1	7.9
Manufacturing	2.7	2.8	1.4	3.5
Mining	1.8	4.1	−6.9	−1.7
Construction	−0.2	0.8	−2.1	−2.1
Services production	1.9	2.5	0.7	0.7
Transportation	1.7	2.4	1.5	−0.4
Communications	5.4	6.0	4.3	4.1
Public utilities	4.4	6.3	0.3	1.2
Trade	2.4	2.9	0.8	2.0
Finance and insurance	0.9	1.4	−0.1	0.2
Real estate	1.2	2.0	−0.2	−0.6
Services	0.9	1.3	0.2	0.1

SOURCE: American Productivity Center.

hours data. Thus, real gross product in those portions of the finance and services groups, as in government, contains no allowance for productivity advance. If productivity grew as much in those industries as in the ones for which independent output estimates were used, productivity in the services sector as a whole would have grown by 0.3 percentage point more over the period as a whole, narrowing the gap with the goods sector as measured. There may, however, be some downward bias in estimates of real product and productivity in the goods sector, particularly in contract construction. So the conclusion still holds that productivity growth in the services sector is significantly below that in goods.

Industry Differences

A glance at Table 1 shows the great differences among the industry groups and their relative importance in terms of persons engaged or gross product originating. This is also true, of course, of the detailed industry components of the groupings.

Within the services sector, the groups differ in the proportions in which they render services to business and to consumers, just as the goods industries produce both final and intermediate products, as detailed in input-output tables. The regulated groups—transportation, electric and gas utilities, and communications (TUC)—are highly capital intensive, while the other groups are generally not so. The other services groups are also quite fragmented into many small establishments. They tend to have higher labor turnover, less well-educated workers (except for professional services), and a higher rate of business failures. All of these characteristics have a bearing on productivity and technological innovation.

With respect to productivity growth rates, shown in Table 3, communications and public utilities groups were among the highest in the economy. Transportation and trade, whose output is measured in terms of the physical volume of goods distributed, had productivity growth rates close to the economy average. It was in the finance and insurance, real estate, and services (FIRES) groups that average rates were well below average, even after allowance was made for the probable downward bias in the output estimates.

Rates of change in labor productivity for a wider range of services industries, prepared by the Bureau of Labor Statistics (BLS), are presented in the chapter by Mark. The finer the industry detail, the greater is the dispersion in rates of change. BLS estimates suggest somewhat better performance of productivity in services than those based on Bureau of Economic Analysis (BEA) real product numbers, which supports the contention that the latter are subject to some downward bias.

Causal Factors Behind Productivity Change

The wide dispersion in rates of productivity change among industries provides a convenient handle for analysis of causal factors. The differential productivity growth rates can be regressed against differences in levels or rates of growth of independent variables believed to have a causal relationship, to identify those that are significant. Results of recent studies have been reported by Kendrick (1986b). Variables that had a significantly large positive association with industry rates of productivity growth in the United States since 1948 are the following:

- ratios of research and development outlays to sales, direct and indirect,
- changes in R&D ratios,
- rates of growth of real fixed capital stocks per worker,
- average education per worker,
- proportion of nonproduction workers and of females in total industry employment,
- variability in layoff rates, and
- economies of scale.

Variables that were negatively associated with productivity changes include:

- the amplitude of cyclical changes in output,
- average hours worked per week and changes in average hours,
- the percentage of workers in an industry belonging to unions,
- days lost because of strikes,

- changes in concentration ratios, and
- changes in the female proportion of the work force.

Since some of the explanatory ratios were correlated (e.g., average education of workers and R&D ratios), only one or the other variable shows a significant (*t* value) in any given multiple regression.

Interindustry analyses help explain the differences in productivity growth between the goods and services sectors, particularly the FIRES group of services (Kendrick, 1987). Stronger and less cyclical output growth was favorable for most services industries. However, these factors were outweighed by others that were less favorable. In particular, the capital/labor ratio grew less rapidly in services; little if any formal R&D was performed directly, and less benefit was derived from the R&D of suppliers of capital and intermediate products. Hours are longer and the length of the standard work week and year has declined less in most services industries than in goods production. The proportions of females and youth, who have less average experience, have generally increased more. Further, concentration is far lower in most services industries than in manufacturing, and the average education of employees is lower. The same is probably true of training, although data are hard to come by concerning this variable.

These findings are generally consistent with analyses of interfirm productivity differences, and with studies of international differences in levels and rates of change in productivity of industries and establishments. A recent study in the latter category also found that exposure to international trade, both exports and imports, and competition from domestic subsidiaries of foreign companies have a positive influence on productivity growth (Davies and Caves, 1987).

At the national level, regression analysis is not a useful analytical tool because of multicollinearity among the time series representing causal forces. The growth accounting approach developed by Edward F. Denison has proved more useful in estimating at least roughly the contributions of various causal factors to macroeconomic growth (Denison, 1985). The technique involves quantifying variables and establishing their relative weights based on their marginal productivities, i.e., their contributions to national income and product. For example, the contributions of various types of investment, tangible and intangible, are evaluated in terms of their rates of return.

Studies by Denison (1985), Maddison (1987), and Kendrick (1979) have identified the following factors as the most important determinants of growth of real gross business product per unit of labor input: (1) the rate of growth of real tangible capital (primarily structures and equipment) per worker; (2) advances in technological knowledge as applied to the processes of production; (3) human capital (primarily education and training) per worker; (4) volume factors: economies of scale associated with growth rates and changes in rates of utilization of capacity associated with cyclical variations; (5) changes in labor efficiency relative to sustainable norms; and (6) governmental measures not reflected in the other proximate determinants, particularly regulations that increase real costs (inputs) but not measured outputs. It is with respect to these categories of causal factors that policy options are discussed in the next section.

It may be noted that the results of the growth accounting studies are generally consistent with the findings of the interindustry analyses. The latter, however, are able to include certain institutional factors, such as levels and changes in concentration of firms and the degree of unionization of workers, which differ by industry but may not exhibit pronounced trends at the national level.

There are other causal forces that are less important at present or are not amenable to policy measures. An example of the former is changes in the composition of output toward industries with higher rates of return. For example, the relative shift out of agriculture until the early 1970s added approximately 0.3 percentage point to productivity growth, according to Denison (1985), but has little effect at present. Also, the decline in the quality of natural resources subtracts about 0.1 percentage point, but there is little that can be done about it. Similarly, the decline in average experience of the work force, associated with the youth bulge beginning in the 1960s, and the increase in the female proportion are not susceptible to policy. In any case, the influence of the changes in age composition of the labor force has been mildly positive in the 1980s as the baby boom generation entered the prime working ages.

The factors discussed above are proximate determinants of productivity change. Underlying these are the basic values and institutions of a society. Factors such as material aspirations, the work ethic, and the efficacy of the profit motive change slowly. The institutional framework of the economy continues to evolve, but it is usually far from clear what the effects of given changes will be. Consequently, issues relating to social and economic reforms are not discussed, but possible changes in tax laws or government regulations, which can have fairly immediate effects on the economy, are considered.

Growth accounting has been applied to explain differences in rates of productivity growth among countries. Denison found that between 1950 and 1962, the chief causes of the faster growth in other Organization for Economic Cooperation and Development countries were

- economies of scale from expansions of local and international markets,
- improved allocation of resources,
- more intensive utilization of capacity, and
- increases in capital per unit of labor.

Education per worker did not increase as rapidly in any of the other countries as it did in the United States. Surprisingly, the technological gap with the United States did not decline significantly in the European countries except for France and possibly Italy.

A follow-up study, based on Denison's model, indicated that between 1960 and 1979 a dramatic narrowing of the technological gap did occur, which reflected technological catch-up coupled with significant increases in R&D spending abroad (Kendrick, 1981). This was associated with stronger increases in capital goods per unit of labor than in the United States as a result of higher savings rates and the opportunities for profitable investments opened up by technological progress and expanding markets.

Growth accounting has also been used to explain changes in growth rates. Thus, the productivity slowdown in the United States between 1973 and 1981, which is also evident in other industrialized countries to varying degrees, was due primarily to the following developments:

- slower growth of real capital per unit of labor, as a result of accelerated labor force growth and a reduced rate of capital formation;
- a decline in advances of technological knowledge consequent on declines in the ratio of R&D to GNP;
- less favorable effects of resource reallocations;
- lesser economies of scale with deceleration of output growth;
- a reduction in rates of utilization of capacity; and
- increasing real costs of compliance with government regulations.

The oil shocks of 1973 and 1979 were major factors behind some of these negative developments.

Since 1981, a reversal of most of the depressing influences has resulted in a modest recovery of the productivity growth rate. However, the recovery has been strong mainly in the manufacturing and agricultural industry groups. Productivity growth continued to be slow in the services sector where a lesser degree of disinflation was associated with a slower growth of output than in the goods sector through 1987.

POLICIES TO PROMOTE PRODUCTIVITY AND TECHNOLOGICAL ADVANCE

The policy options discussed in this section are general and not industry or technology specific. The focus is on measures that can be taken mainly by the federal government to stimulate advances in technological knowledge and their application to the ways and means of production in the directions judged to be most productive by the players in the enterprise economy.

A lengthy paper "Policies to Promote Productivity Growth" (Kendrick, 1980) contained 99 policy options. A good many of the measures have since been adopted, and others proposed. A number of measures are now suggested under six major headings that correspond to the major causal forces behind productivity advance: economic expansion, increasing technological knowledge, savings and tangible investments, human investment, labor efficiency, and government policy not covered in the previous points.

Volume Factors   Rates of change in output and productivity show a strong positive correlation, reflecting economies of both scale and utilization. Economies of scale are a function of the rate of growth of output, since growth opens up opportunities for greater specialization of people, plants, and equipment, and spreads overhead functions over more units. Cyclically, expansion from lower rates of utilization of capacity toward most efficient rates boosts productivity and vice versa. In general, innovation is easier when output is expanding smartly than when it stagnates or declines.

The economic expansion beginning in late 1982 proceeded rapidly through mid-1984, then slowed drastically through 1986. Rates of utilization of industrial capacity declined, and so did the rate of productivity growth, especially in services. It would be desirable if demand in the economy increased at least as fast as the expansion of productive capacity, estimated at 3–4 percent per year. Marked cyclicality of demand also slows the growth of capacity and productivity. The slowdown for the decade beginning in 1973 was due in part to the most severe contractions since World War II, which took place in 1974 and 1982.

The objective of relatively steady growth of real GNP around the "natural" rate of unemployment requires appropriate macroeconomic policies. Since the unemployment rate in 1987 averaged 6.2 percent—approximately 1 percentage point more than the natural rate below which wage and price inflation tends to accelerate—growth could probably proceed at about 4 percent for a couple of years without seriously heating up inflation. Then growth should slow somewhat, reflecting the slower growth of the labor force projected for the 1990s and the assumed leveling of unemployment at 5–5½ percent of the civilian labor force.

An important part of the stronger expansion should be a continuation of the faster growth of exports than of imports in real terms that started in the fall of 1986, due in part to the major decline in the foreign exchange value of the dollar since early 1985. To promote continued significant declines in the trade balance, U.S. foreign economic policy must not only seek to maintain an equilibrium value of the dollar; it must also encourage stronger economic growth of our major trading partners and continue negotiations for freer, fairer trading rules and practices on their part and ours. It should be noted that exports of services as well as of goods have benefited from the improved trade balance.

More generally, the expansion of the share of the services sector in the economy has tended to reduce cyclical fluctuations. This makes the task of macroeconomic policy easier. On the other hand, the shift to services tends to reduce the growth of productivity and productive capacity, which underlines the importance of seeking measures to accelerate productivity growth in services.

Promoting Technological Knowledge   In the modern era, the fountainhead of scientific knowledge and technological advance is research and development. Ratios of R&D, direct and indirect, to sales within manufacturing are significantly correlated with rates of productivity advance. The great increase in the ratio of national R&D to GNP from a small fraction in 1919 to approximately 3 percent in the mid-1960s was a major reason for the acceleration in productivity growth after World War I. The subsequent decline in the ratio until the latter 1970s is cited as a reason for the productivity slowdown after 1973, given the lags between R&D and commercial effects. The increase in both business- and government-financed R&D since the late 1970s contributed to the pickup in productivity growth after 1982, particularly in manufacturing.

It is important that R&D continue to increase at least in line with GNP, and preferably somewhat faster if the supply of scientists and engineers permits without undue inflation of salaries. Most of the increase in government-financed R&D has been for national security purposes. As that increase slackens, it will be desirable that financing of applied R&D for civilian purposes be increased, particularly in areas that would benefit production of services. Recent increases in funding of basic research and the planned doubling of appropriation requests for the National Science Foundation (NSF) over the next five years are welcome developments.

The passage of a 25 percent incremental R&D tax credit in 1981 helped maintain the strong increases in business R&D. This was extended in 1986, but at a 20 percent rate, and the extension was for a temporary, three-year period. This may have contributed to a slowing of increases in business R&D. It is recommended that the credit be increased to at least 25 percent, that the definition of R&D be broadened, that increments be computed on a stable base, and that the credit be made permanent because R&D programs require long-term planning (Brown, 1984).

Patent, copyright, trademark, and trade secret protection has been strengthened pursuant to recommendations in the 1985 report of the President's Commission on Industrial Competitiveness. The protection of intellectual property has been raised to a priority concern in the coming round of multilateral trade talks. Some people recommend lengthening the period of patent protection from 17 to 20 years, which would help compensate for delays in commercialization of inventions because of federal regulations. There is also a need to strengthen the Patent and Trademark Office to improve search capabilities and the reliability of patent grants.

Serious consideration should be given to proposals in the President's Competitiveness Initiative based on the recommendations of his commission (Council on Competitiveness, 1987). Among the initiatives are the following:

- establishment of new university-based, interdisciplinary science and technology centers to perform research in areas that contribute to U.S. competitiveness (including services);
- creation of a technology share program involving multiyear, joint basic and applied research with consortia of U.S. firms and universities;
- initiation of an exchange program between scientists and engineers in the public and private sectors;
- improved industry access to federal science and technology efforts to increase transfers and accelerate the spin-off of defense technologies to the private sectors;
- programs to increase scientific literacy; and
- grants to update the equipment in academic research laboratories, 20 percent of which is obsolete.

The National Cooperative Research Act of 1984 removed antitrust barriers to joint research. More than 40 consortia have notified the Department of Justice and the Federal Trade Commission that they intend to take advantage of the new opportunities. It is important that companies in the services industries be fully represented in these and related initiatives.

Savings and Investment   There is a significant positive correlation between rates of change in real capital per worker and output per worker. This is not just a matter of the quantity of capital goods, but also of their quality, because new plant and equipment are carriers of cost-reducing technological innovations. The volume of business investment depends on expected rates of return and on the cost of capital, particularly interest rates, which interact with savings rates. Savings constitute ultimate constraint on the volume of investment that can be undertaken.

A negative aspect of income taxation is that it drives a wedge between the returns earned by investment and the income accruing to investors. It also drives a wedge between market rates of interest and the interest received by savers. Another way of expressing the negative influence of income taxation on savings and investment is that it represents double taxation of savings in that income is taxed and so is the income from the investments into which savings flow.

From this point of view, the reductions in both corporate and personal income tax rates in the tax acts of 1981–1982 and 1986 have had a positive effect on savings and investment. On the other hand, the elimination of the investment tax credit and the accelerated cost recovery system in the 1986 act resulted in a drop of business investment between the last quarter of 1985 and the first quarter of 1987, contributing to sluggish economic growth. Consideration should be given by Congress to restoration of an investment tax credit.

More fundamentally, it will be desirable to further de-emphasize income taxation, or eliminate it, and to substitute consumption-based taxes. Economists at the Brookings Institution have made a good case for a consumed income tax (Bosworth, 1984). A value-added tax is also neutral with respect to savings and consumption decisions, but appears to be politically unpopular.

An unfortunate aspect of the Tax Act of 1986 was the increase in capital gains taxation beginning in 1987. Business investments and the formation of new businesses are undertaken in part in hope of realizing capital gains in the future. Capital gains taxation reduces that incentive. One attractive proposal for reducing capital gains tax rates is to reduce the capital gains on which taxes are levied to the extent that the price level (as measured by the consumer price index) had risen between the dates of purchase and sale of the assets.

With respect to interest rates, the most effective way to hold them down or reduce them further at this time is through continued reductions in the federal deficit. Government deficits absorb funds that would otherwise be available for private investment, and competition between business and governments in the money markets keeps interest rates higher than they would be if budgets were balanced. Monetary policy should not be overly stimulative, however, because of the inflationary potential, but it can at least be neutral in view of the remaining slack in the economy at this time.

Human Investments   Growth accounting studies show that increases in average education and training of the work force contributed about 0.7 percentage point to the 2.2 percent increase in labor productivity between 1948 and 1986 (Denison, 1985; Kendrick, 1979). On a cross-sectional basis, as noted earlier, there is a significant positive correlation between rates of change in average education and output per labor hour by industry. Along with tangible investments and technological advances, education and training rank among the most important forces promoting productivity advance. They are basic to producing the scientists, engineers, and business managers responsible for innovation and to preparing the labor force generally to operate increasingly complex technology.

It is important that both public and private outlays for education and retraining continue to increase as a ratio to GNP. The 1985 report of the President's Commission on Industrial Competitiveness made several recommendations in the human resources area. Among these were:

- expansion of federal training and assistance programs for displaced workers;

- greater funding of engineering education, including expansion of NSF engineering centers;
- sustained federal support for a program of basic and prototype research in educational software through NSF and the Department of Labor; and
- more effective dialogue among government, industry, labor, academia, and other interested parties to develop consensus on educational and other measures to increase competitiveness.

A task force headed by the Secretary of Labor concluded that government programs to deal with displaced workers should be consolidated and funding for retraining nearly tripled. The President's fiscal year 1988 budget provided almost $1 billion for worker adjustment assistance, well above the previous year. The President's Competitiveness Initiative recommends an $800 million program under the Job Training Partnership Act to provide summer jobs, remedial education, and skills training for disadvantaged youths.

The administration's 1988 budget requested an increase of $600 million for an experimental loan program to allow students to borrow more money than currently permitted. However, this would come at the expense of proposed reductions in vocational training and in funding of higher education.

An older proposal is for tuition tax credits. Another is for development and diffusion of new educational methods and technologies by the National Institutes of Education, along with research to determine those that are most effective in facilitating learning. Increasing emphasis should be placed on lifelong learning. The National Academy of Engineering (1986; 1988) has stressed this with respect to the need for continuing education of engineers.

Health is another form of human investment. Here the emphasis must be on the development of new and more effective medical technology and treatments, and on the continuing education of the public in healthful lifestyles. Further increases in the productivity of health care facilities are essential to bring down their relative costs.

Labor Efficiency With a given level of technological knowledge, the actual output of individuals and organizations depends on the degree of efficiency relative to a sustainable optimum. In the short run, the productivity of most individuals and groups can be increased significantly with proper leadership, motivation, and inputs of brain and brawn.

Over the past decade, there has been a major shift in management philosophy and practice in many firms and other organizations from the old "Theory X" military chain of command model to "Theory Y," involving more participative management and the use of employee involvement (EI) systems. Some of the EI plans, such as quality circles and joint labor-management productivity teams, do not involve special financial incentives. Others, such as Scanlon or Rucker plans, Improshare, productivity gain-sharing and profit-sharing plans, do incorporate financial incentives with rewards linked to performance. A number of surveys of EI systems, both with and without financial incentives, indicate that managers believe they have significantly enhanced productivity.[2]

Installation and maintenance of EI plans, and systems for tying pay to performance, depend on management initiatives, of course. However, both the Department of Labor and the Federal Mediation and Conciliation Service have staffs that can assist in developing cooperative efforts of management and labor to promote productivity. They can also assist in union-management negotiations involving the moderation or elimination of restrictive work rules or jurisdictional conflicts that impair productivity. The longer term payoff of EI plans comes when they elicit worker suggestions for cost-reducing innovations in the workplace.

An executive order issued by the Office of Management and Budget in February 1986 requires federal agencies and departments to develop productivity measures, to set up systems for involving workers in efforts to improve productivity, and to use productivity improvement goals in budgeting. These efforts are being reviewed and evaluated by the General Accounting Office, which is in a position to recommend modifications to make them more effective. The federal government is prepared to help state and local governments in similar efforts. Getting "more bang for the tax buck," or spending fewer bucks for the bang, in government is a significant element in raising national productivity.

Governmental Policies   In providing the legal framework within which private enterprises operate, government plays an important role in the productivity of the business economy. The aggregates and composition of government expenditures and revenues also have a significant impact on the private economy. The effects of taxation and the government surplus or deficit on private savings and investment have been discussed. Likewise, the proportions of government expenditures devoted to public investments in infrastructure and in intangibles such as R&D, education, training, and health affect the productivity of the private economy. Many economists think that a separate capital budget for government would help focus attention on the importance of public investment. Certainly, strict cost-benefit analyses should be applied in selecting specific investment projects to be undertaken.

Of particular concern to business has been the proliferation of governmental regulations since 1970. Because the costs of complying with regulations increase inputs while the presumed benefits do not increase output as measured, social regulations are believed to have played a part in the productivity slowdown. It is important that the administration persevere in its efforts to rationalize the regulatory process, requiring early public involvement before promulgation of new regulations, coordinating agency activities, tightening procedural requirements for evaluative choices of proposed regulations and reevaluation of existing regulations, and improving evaluation methods. Where regulations are found to have negative impacts, provisions for minimizing or eliminating such impacts should be made. Agencies should be held accountable for the claimed benefits relative to costs and risks. Performance should be reviewed regularly, and failure to achieve satisfactory results would be cause for deregulation or substitution of alternative procedures.

Regulations should be goal oriented, rather than specifying means. Agencies should try to compensate for the greater relative burdens placed on small firms by regulations. They should study and take account of the impact of their actions on innovation, productivity, and costs, and thus the international competitiveness of the U.S. economy.

Reducing or eliminating economic regulations in a variety of industries has had salutary effects, including increases in productivity (Bailey, 1986). Technological advances have increased competition within and among industries, and it is to be hoped that the movement toward reducing or eliminating economic regulations will continue.

In recent years, antitrust laws have been interpreted more broadly by the Department of Justice and the Federal Trade Commission to permit mergers that would increase efficiency without restraining trade. This contributes to productivity growth. However, it remains essential that mergers and collusive activity of firms which would lead to monopolistic pricing be prohibited. Healthy competition is a powerful force promoting innovation, productivity advance, and reduction of real costs per unit of output. The effective functioning of the U.S. economy depends on its maintenance.

CONCLUSION

During the golden quarter century (1948–1973) after World War II, productivity trends in major portions of the services sector were significantly below the national average. It is a particular cause for concern that since the 1973–1981 productivity slowdown which affected all industries, productivity growth in services failed to rebound (except for trade) as it did in goods production other than construction. Because output and productivity in finance and services industries are understated, it is important that statistical agencies devote more attention to improving the measures. Even after allowance for understatement, however, it seems clear that productivity growth in most services groups, other than communications and trade, has been significantly below the U.S. business sector average.

Most of the policy options discussed above would help to improve productivity in the business economy generally, but since most services industries are much more labor intensive than the goods sector, priority should initially be given to measures that improve the quality and efficiency of labor, including managers. Schools of business should increasingly slant their curricula toward the probability that their graduates will pursue careers in services establishments, now that manufacturing accounts for only 18 percent of total employment in the United States. Improvements in the quality of public education and the reduction of dropout rates are of special value to services industries. The Job Training Partnership Act and other formal training programs should stress training for services jobs. Companies in the services sector should try to increase the amount of training they can offer employees. More widespread adoption of employee involvement plans and the use of incentive pay systems could have a significant effect on efficiency as discussed earlier. In those services industries in which difficulties in measuring work units would militate against productivity gain sharing, profit-sharing schemes offer an attractive alternative.

With respect to technological advance, firms in the service industries should take full advantage of the present opportunities to form research consortia free of the threat of antitrust prosecution. New university-based interdisciplinary science and technology centers must give full weight to the important role of services in contributing to U.S. competitiveness in international economic relations. Given the fragmented nature of many services industries, trade associations in these industries should move to sponsoring more research, development, and engineering activity and should apply for grants from the NSF and other interested federal agencies.

Economists generally hold that it it not desirable to give special investment incentives to selected sectors or industries. These tend to distort the allocation of investment funds based on the best judgments of financial institutions regarding relative profit prospects. The best stimulus to investment comes from steady, noninflationary economic expansion in line with the growth of productive potential of the economy as discussed earlier. The decline of the dollar since October 19, 1987 should help promote further reductions in the negative trade balance. If additional stimulus to growth is needed, resumption of the 10 percent investment tax credit would help promote increases in real product and productivity across the board. Effective marketing strategies of capital goods manufacturers involving education of managers—especially of smaller firms—in applying the latest technologies, are helpful to services industries. Beyond this, continuing efforts of the Small Business Administration to aid small firms in meeting regulatory requirements, entering export markets, gaining access to financing on reasonable terms, etc., will be of special benefit to services industries.

The biggest factor that most services industries have working for them is something unaffected by policy, i.e., generally high income elasticities and low price elasticities of demand. This ensures the continued expansion of the service sector with attendant economies of scale. Relatively strong expansion provides a favorable environment for managements of services firms to meet their greatest challenge—technological innovation.

NOTES

1. For publications and information, write the American Productivity Center, 123 North Post Oak Lane, Houston, TX 77024.
2. See New York Stock Exchange (1982). A new survey was conducted in 1987 by the American Productivity Center, Houston.

REFERENCES

Bailey, E. 1986. Price and productivity change following deregulation: The U.S. experience. Economic Journal (March):1–17.

Bosworth, B. P. 1984. Tax Incentives and Economic Growth Washington D.C.: The Brookings Institution.

Brown, K. M. 1984. The R&D Tax Credit, Issues in Tax Policy and Industrial Innovation. Washington, D.C.: American Enterprise Institute for Public Policy Research.

Council on Competitiveness. 1987. America's Competitive Crisis: Confronting the New Reality. Washington, D.C.: Council on Competitiveness (April).

Davies, S., and R. E. Caves. 1987. Britain's Productivity Gap. New York: Cambridge University Press.

Denison, E. F. 1985. Trends in American Economic Growth, 1929–1982. Washington D.C.: The Brookings Institution.

Guile, B. R, and J. B. Quinn. 1988. Managing Innovation: Cases from the Services Industries. Washington, D.C.: National Academy Press.

Kendrick, J. W. 1979. Productivity trends and the recent slowdown: Historical perspective, causal factors, and policy options. Pp. 17–69 in Contemporary Economic Problems, Fellner, ed. Washington, D.C.: American Enterprise Institute for Public Policy Research.

Kendrick, J. W. 1981. International comparisons of recent productivity trends. Pp. 125–170 in Essays in Contemporary Economic Problems, W. Fellner, ed. Washington: American Enterprise Institute for Public Policy Research.

Kendrick, J.W. 1980. Policies to promote productivity growth. Pp. 45–135 in Agenda for Business and Higher Education. Washington, D.C.: American Council on Education.

Kendrick, J. W. 1986a. Output, inputs, and productivity in the service industries. Pp. 60–89 in Statistics About Service Industries: Report of a Conference. Washington, D.C.: National Academy Press.

Kendrick, J. W. 1986b. Differences among industries in productivity growth. Economic Policy Studies Occasional Paper. Washington: American Enterprise Institute for Public Policy Research (November).

Kendrick, J. W. 1987. Service sector productivity. Business Economics (April):18–24.

Maddison, A. 1987. Growth and slowdown in advanced capitalist economies. Journal of Economic Literature 25(2):649–698.

National Academy of Engineering. 1986. Federal Actions for Improving Engineering Research and Education. Washington, D.C.

National Academy of Engineering. 1988. Focus on the Future: A National Action Plan for Career-Long Education for Engineers. Washington, D.C.

New York Stock Exchange. 1982. People and Productivity, A Challenge to Corporate America. New York: New York Stock Exchange Office of Economic Research.

Technology and the Services Sector: America's Hidden Competitive Challenge*

STEPHEN S. ROACH

Information technology has always played an important role in the services sector of the U.S. economy. In recent years, however, services industries have stepped up their acquisitions of computers, telecommunications equipment, and other such products dramatically. As a result, the broad segment of the economy that can be classified as services providers now owns about 84 percent of the total U.S. stock of information technology items. Moreover, relative to goods-producing industries, a much larger proportion of the services sector's capital budgets is spent on information technology, revealing a significantly greater dependence by services on such technology as a factor of production. This reliance underscores information technology's strategic importance in the U.S. competitive challenge: with services now the predominant mode of economic activity in the United States, a productivity payback from information technology is absolutely essential to keep the economy on a longer term path of sustainable growth.

So far, the services sector has little to show for its spending binge on technology. Quite simply, massive investments in information technology have failed to boost national productivity growth in the present decade. Furthermore, with manufacturing productivity now on the rebound, problems in the services sector loom increasingly large in America's broader competitive struggle. It is certainly not too late. New and creative applications of information technology could still enhance the productivity performance of the services sector's predominantly white-collar work force. Until that

---

* Some of the findings reported in this chaper are drawn from studies previously published by Morgan Stanley; for the most recent such effort, see Morgan Stanley (1987).

payback begins to occur, however, the role of technology spending will be under growing suspicion (Bowen, 1986; Neikirk, 1987; Schneider, 1987).

In what follows, an attempt is made to provide a detailed industry-by-industry assessment of services sector spending on information technology. By way of background, the broad contours of capital formation in services industries are first examined over the post-World War II era.

CAPITAL ACCUMULATION IN THE SERVICES SECTOR

Over the course of contemporary economic history, the services sector has emerged as a major force in shaping capital formation in the United States. As the upper portion of Figure 1 indicates, up until about the mid-1960s capital budgets in the goods-producing and services-providing sectors were of roughly equal magnitudes in "real," or inflation-adjusted, terms. Over the past 20 years, however, a sizable gap has opened between the investment programs of these two segments of the economy; at present, services-providing industries account for nearly 60 percent of corporate America's total

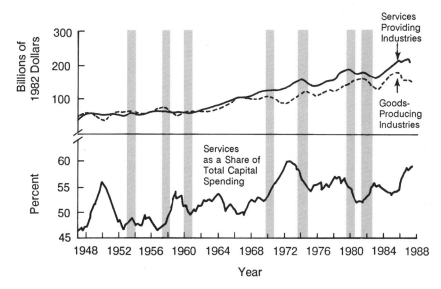

FIGURE 1   Post-World War II trends in services sector capital spending.
NOTE: Shaded areas indicate recessionary periods as designated by the National Bureau of Economic Research. The goods-producing sector is defined to include manufacturing, mining, and construction; the services sector—from which government entities are excluded—consists of trade, communications, transportation, public utilities, finance, insurance, real estate, business services, professional services, health, and legal and educational institutions.

outlays on plant and equipment. Moreover, as the lower portion of this figure illustrates, services sector investment budgets, in general, have tended to be relatively immune to the ups and downs of the business cycle. With one exception—the contraction of 1973 to 1975—the investment share of this segment of the economy actually has risen during the major recessions of the past 35 years. This does not mean that services providers get away unscathed during recessions. However, as Figure 1 shows, investment cutbacks during periods of cyclical distress are generally smaller for these industries than for manufacturers, with the latter still bearing the brunt of most serious shortfalls in economic activity.

Table 1 provides a summary of investment trends in major segments of the services sector over the past three and half decades. In the 1950s and 1960s, capital investment of services providers grew at or slightly below spending rates in goods-producing industries. Not until the 1970s did services really come to the fore as the leader of capital formation in the United States. During the 1970s decade, services providers increased capital budgets (in real terms) at a 4.6 percent average annual rate; although that performance represented a slowdown from the prior decade, it was one-third faster than the 3.4 percent yearly gains in the goods-producing segment. In the 1980s, overall investment advances have slowed further, although gains in the services segment have continued to outdistance those in goods-producing industries by a considerable margin. By industry, the recent strength in services sector investment has been concentrated in finance, insurance, and real estate; wholesale trade; and retail trade segments of the economy. By contrast, over the past decade and a half, capital spending growth has slowed for trans-

TABLE 1   Shifting Trends in Business Fixed Investment, Average Annual Growth Rates (percent)

Business Sector	1950s	1960s	1970s	1980s[a]
ALL INDUSTRIES	2.8	6.9	4.0	2.2
Goods producing	3.6	6.9	3.4	1.8
Manufacturing	3.9	7.6	2.7	2.6
Other	2.0	0.5	11.4	−3.6
Services providing	2.3	6.8	4.6	2.6
Transportation	−0.4	6.6	0.8	−2.0
Communications	7.2	8.9	4.9	1.1
Public utilities	0.9	6.8	3.6	−0.5
Trade	4.8	4.5	7.4	6.0
Finance and insurance	10.8	8.8	11.0	12.8
Personal and business services	4.5	8.2	4.2	−2.4

[a]Through 1985.

NOTE: Figures are based on constant 1982 dollars and are taken from the plant and equipment survey of the U.S. Department of Commerce.

portation services, public utilities, personal and business services, and communications.

Similar conclusions are apparent in examining recent trends in the services sector's stock of information technology capital, which is hardly surprising since such a stock reflects the accumulation of past investment flows. Figure 2 summarizes three and a half decades of growth in the U.S. business sector's capital stock, with an emphasis on the diverging experience of services and goods producers. While capital stock growth in both segments has slowed in the present decade relative to peak increases in the 1960s, gains in the services sector have generally held up much better, running "only" 25 percent below the record rates of two decades ago, compared to a 75 percent growth shortfall in manufacturing.

Overall trends in services sector capital accumulation mask some important differences among industries, as Table 2 illustrates. For trade establishments, capital stock growth has actually picked up in the present decade relative to the pace of the 1970s. A similar trend is observable for a collection of services sector industries referred to as "information producers"—communications, finance, insurance, real estate, business services, and legal services. All in

FIGURE 2   Growth in the U.S. capital stock.
NOTE: Figures are average annual growth rates based on constant 1982 dollars; capital stock data are net stocks using constant-cost valuation and straight-line depreciation patterns; the data are taken from U.S. Department of Commerce statistics (described in Musgrave, 1986). The "1980s" includes data through 1985. For our analysis, an extension of the industry detail reported in that article was used, which provides both an industry and a commodity breakdown of the private nonresidential capital stock. The resulting industry-commodity capital stock matrix contains 59 industry groupings; each industry, in turn, has up to 21 categories of equipment and structures. We are grateful to Mr. Musgrave for both providing these data and being most generous with his time in clarifying several issues regarding data construction.

TABLE 2   Postwar Trends in the Capital Stock, Average Annual Growth Rates (percent)

Business Sector	1950s	1960s	1970s	1980s[a]
ALL INDUSTRIES	3.8	4.3	3.6	3.1
Goods producing	4.3	3.8	3.2	2.1
Manufacturing	3.5	4.7	3.3	1.7
Other	5.7	2.2	3.1	2.8
Services providing	3.6	4.5	3.8	3.6
Information producers	6.4	6.8	4.6	5.0
Communications	6.4	6.7	6.0	4.6
Finance	5.8	8.7	7.4	7.8
Insurance	7.2	6.2	3.4	6.3
Real estate	6.2	6.2	3.1	4.0
Business services	10.3	11.4	6.8	7.9
Legal services	5.0	2.8	1.4	6.2
Trade	3.9	6.8	5.1	5.6
Other	2.4	2.7	2.8	1.5

[a]Through 1985.

NOTE· Figures are based on constant 1982 dollars and are derived from U.S. Department of Commerce statistics.

all, whether it is investment spending or capital stock growth, the message that comes through loud and clear is that the services sector clearly has taken over as the engine of capital formation in the United States.

TECHNOLOGY AND SERVICES SECTOR INVESTMENT

Not surprisingly, it turns out that the services sector's impressive record on capital accumulation largely reflects a voracious appetite for information technology. Figure 3 divides services sector capital stock growth into technology and "nontechnology" items. As the middle panel shows, the stock of information technology capital in the services sector has expanded at close to a 10 percent average annual rate over the past 35 years; although the growth rate slipped a bit in the 1960s and 1970s, it has accelerated significantly in the present decade to approximately a 12 percent average annual pace—a record performance by historical standards. Indeed, had it not been for the rapid gains in such technology spending, total services sector capital accumulation would have slowed considerably further over the past decade and a half. As Figure 3 shows, thus far in the 1980s, nontechnology services sector capital has risen at only a 2.5 percent yearly pace, over one-third slower than average gains of the period 1950–1979 and the weakest performance of any decade in the postwar era.

Figure 4 offers more insight into shifting trends in services sector invest-

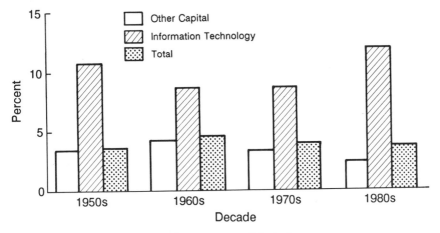

FIGURE 3 Growth in the services sector capital stock.
NOTE: Figures are average annual growth rates based on constant 1982 dollars; capital stock data are net stocks using constant-cost valuation and straight-line depreciation patterns. The "1980s" includes data through 1985. Information technology capital, or information-processing equipment, is defined to include computers, office and accounting machinery, communications equipment, instruments, and photocopiers and related equipment. Industry estimates of information technology capital were obtained by a technology "sort" of the industry-commodity capital stock matrix described for Figure 1. Obviously, other nonservices industries have increased their reliance on new technologies, ranging from machine tools and motor vehicles to biogenetics and artificial intelligence. These items are not included in this analysis of information technology capital largely because of the relatively small role they play in overall business activity and also because they are not separately identifiable in the measurement framework employed, which is based on gross national product.

ment from 1970 to 1985; the functional composition of the capital stock in this segment of the economy is shown. Of the four major product categories identified, only information technology has experienced an increase in its share of the total services sector's capital stock; in 1970 information technology accounted for only 6.4 percent of the total services capital, whereas 15 years later the share had risen almost 2½ times to 15.5 percent.

This sharp surge in information technology purchases has been accommodated by a significant rearrangement in the capital budgets of services-providing industries. Most significant in this regard has been a contraction of the portion of capital now invested in buildings. The transportation equipment share has dropped only fractionally, and industrial equipment has essentially maintained its relatively small portion of 15 years earlier.

For comparative purposes, Figure 5 provides a parallel analysis of compositional shifts in the capital stock of the goods-producing segment of the

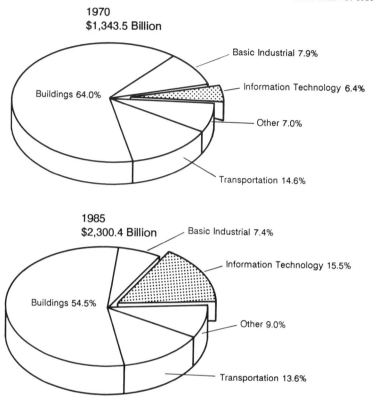

FIGURE 4   Shifting composition of services sector capital stock.
NOTE: Figures are based on constant 1982 dollars. Because of rounding, totals may not add up to 100.0 percent. Information technology capital includes office, computing, and accounting machinery; communications equipment; instruments; and photocopiers and related equipment. In addition to the information technology grouping, this functional breakdown also includes the following broad categories: "basic industrial capital," which encompasses fabricated metal products, engines and turbines, metalworking machinery, special industry machinery, general transmission, distribution, and industrial apparatus; "buildings," which consists of commercial, office, and public utility structures; and "transportation," which is composed of automobiles, trucks, trailers, buses, commercial aircraft, and rail equipment.

U.S. economy. While there have been analogous rearrangements of productive capital, goods producers remain considerably less dependent on information technology as a factor of production. In 1985, approximately 6 percent of their total capital was invested in such technology items. While that is about three times the average share of the 1970s, it is less than half the portion currently found in the services sector. Without question, the most

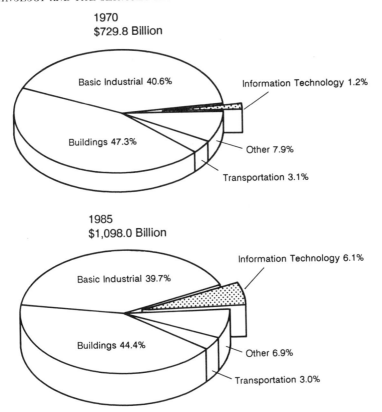

FIGURE 5 Shifting composition of goods-producing sector capital stock.
NOTE: Figures are based on constant 1982 dollars. Because of rounding, totals may not add up to 100.0 percent. Information technology capital includes office, computing, and accounting machinery; communications equipment; instruments; and photocopiers and related equipment.

intensive users of information technology are in the services sector; moreover, this same segment has continued to move dramatically to increase this dependence.

INDUSTRY DETAIL

Tables 3 through 5 provide a detailed industry-by-industry assessment of shifting trends in the stock of information technology capital in the services sector. Benchmark assessments describing the concentration of technology ownership within various segments of the services grouping can be found in Tables 3 and 4. In 1985—the last year for which complete industry and

TABLE 3   Where is America's Information Technology Capital? (billions of constant 1982 dollars)

Business Sector	1950s	1960s	1970s	1985
ALL INDUSTRIES	28.8	61.5	142.7	423.6
Goods producing	4.7	6.4	17.2	66.7
Services providing	24.1	55.1	125.5	356.9
Transportation	1.0	1.2	1.5	3.1
Rail	0.4	0.6	0.8	0.6
Nonrail	0.5	0.6	0.7	2.4
Air	0.1	0.1	0.1	1.2
Trucking	0.1	0.1	0.0	0.2
Other	0.4	0.4	0.5	1.0
Communications	10.5	30.6	73.4	159.7
Telephone and telegraph	10.4	29.9	71.3	154.4
Broadcasting	0.2	0.7	2.1	5.3
Public utilities	0.8	1.4	3.9	13.4
Electric	0.6	1.1	3.4	11.0
Gas and other	0.2	0.3	0.5	2.5
Total trade	0.5	1.1	5.7	41.1
Wholesale trade	0.3	0.7	4.0	33.9
Retail trade	0.2	0.4	1.6	7.2
Finance, insurance, and real estate	8.1	13.7	24.6	90.3
Finance and insurance	0.8	1.3	4.2	46.4
Banks (including Federal Reserve)	0.2	0.4	1.6	20.5
Credit agencies	0.1	0.3	1.1	15.6
Securities brokers	0.0	0.0	0.1	0.9
Insurance carriers	0.2	0.2	0.7	6.1
Insurance agents	0.2	0.3	0.3	1.0
Investment holding companies	0.1	0.1	0.4	2.3
Real estate	7.3	12.4	20.4	44.0
Services	3.1	7.1	16.4	49.3
Hotels and lodging	0.0	0.0	0.1	0.9
Personal	0.1	0.8	1.6	1.5
Business	0.3	1.2	4.3	22.3
Auto repair	0.1	0.1	0.1	0.8
Miscellaneous repair	0.0	0.0	0.0	0.2
Motion pictures	0.1	0.4	1.4	2.4
Amusement and recreation	0.4	1.1	1.9	3.8
Health	1.2	2.4	5.4	13.7
Legal	0.0	0.1	0.1	0.7
Educational	0.1	0.1	0.2	0.7
Other	0.7	1.0	1.3	2.1

NOTE: Figures are averages over designated intervals and are Morgan Stanley estimates derived from the Industry-Commodity Capital Stock Matrix of the U.S. Department of Commerce.

TABLE 4   Industry Shares of Information Technology Capital

Business Sector	Percentage				Change: 1985 vs. 1970s	
	1950s	1960s	1970s	1985	Percentage Points	Ratio[a]
ALL INDUSTRIES	100.0	100.0	100.0	100.0	0.0	1.0
Goods producing	17.1	10.6	11.4	15.8	4.3	1.4
Services providing	82.9	89.4	88.6	84.2	−4.3	1.0
Transportation	3.6	2.0	1.1	0.7	−0.4	0.7
Rail	1.6	1.0	0.6	0.1	−0.5	0.2
Nonrail	2.0	1.0	0.5	0.6	0.1	1.1
Air	0.3	0.2	0.1	0.3	0.2	3.3
Trucking	0.3	0.2	0.0	0.0	0.0	1.7
Other	1.4	0.7	0.4	0.2	−0.1	0.6
Communications	36.1	48.8	52.0	37.7	−14.3	0.7
Telephone and telegraph	35.5	47.8	50.5	36.5	−14.0	0.7
Broadcasting	0.6	1.1	1.5	1.2	−0.3	0.8
Public utilities	3.0	2.2	2.7	3.2	0.5	1.2
Electric	2.2	1.7	2.3	2.6	0.3	1.1
Gas and other	0.9	0.5	0.3	0.6	0.2	1.7
Total trade	1.8	1.8	3.7	9.7	6.0	2.6
Wholesale trade	1.1	1.1	2.6	8.0	5.4	3.1
Retail trade	0.8	0.7	1.1	1.7	0.6	1.5
Finance, insurance, and real estate	27.8	23.0	17.5	21.3	3.8	1.2
Finance and insurance	2.7	2.3	2.7	10.9	8.2	4.0
Banks (including Federal Reserve)	0.6	0.6	1.0	4.8	3.8	4.8
Credit agencies	0.5	0.5	0.7	3.7	3.0	5.5
Securities brokers	0.1	0.1	0.1	0.2	0.1	2.1
Insurance carriers	0.6	0.4	0.5	1.4	1.0	3.1
Insurance agents	0.6	0.5	0.2	0.2	0.0	1.2
Investment holding companies	0.4	0.2	0.3	0.5	0.3	1.8
Real estate	25.1	20.7	14.8	10.4	−4.4	0.7
Services	10.5	11.6	11.7	11.6	0.0	1.0
Hotels and lodging	0.0	0.0	0.0	0.2	0.2	5.9
Personal	0.5	1.2	1.2	0.4	−0.8	0.3
Business	1.1	1.9	3.0	5.3	2.2	1.7
Auto repair	0.2	0.1	0.0	0.2	0.1	4.5
Miscellaneous repair	0.0	0.0	0.0	0.0	0.0	2.6
Motion pictures	0.5	0.6	1.0	0.6	−0.4	0.6
Amusement and recreation	1.4	1.8	1.4	0.9	−0.5	0.7
Health	4.1	3.9	3.8	3.2	−0.6	0.8
Legal	0.1	0.1	0.1	0.2	0.1	1.6
Educational	0.3	0.2	0.1	0.2	0.1	1.7
Other	2.3	1.7	0.9	0.5	−0.4	0.6

[a]1985/1970s average.

NOTE: Figures are averages over designated intervals and are Morgan Stanley estimates derived from the Industry-Commodity Capital Stock Matrix of the U.S. Department of Commerce.

TABLE 5 Information Technology Capital as a Share of Each Industry's Overall Capital Stock

	Percentage				Change: 1985 vs. 1970s	
Business Sector	1950s	1960s	1970s	1985	Percentage Points	Ratio[a]
ALL INDUSTRIES	2.5	3.7	5.8	12.5	6.7	2.2
Goods producing	1.1	1.1	2.0	6.1	4.1	3.0
Services providing	3.3	5.2	7.7	15.5	7.8	2.0
Transportation	0.4	0.5	0.6	1.1	0.5	1.9
Rail	0.2	0.5	0.7	0.6	−0.1	0.9
Nonrail	0.8	0.7	0.5	1.4	0.9	2.8
Air	1.5	0.7	0.4	3.0	2.5	7.3
Trucking	0.5	0.5	0.1	0.4	0.3	3.7
Other	0.8	0.8	0.8	1.3	0.5	1.7
Communications	20.0	30.6	40.8	53.4	12.6	1.3
Telephone and telegraph	20.8	31.9	42.1	54.5	12.5	1.3
Broadcasting	5.7	10.6	19.9	32.9	13.0	1.7
Public utilities	0.5	0.6	1.1	3.1	2.1	2.9
Electric	0.5	0.6	1.2	3.2	2.0	2.6
Gas and other	0.5	0.4	0.6	2.8	2.3	5.0
Total trade	0.7	0.9	2.5	11.1	8.7	4.5
Wholesale trade	1.6	1.8	4.9	22.5	17.6	4.6
Retail trade	0.4	0.5	1.1	3.3	2.1	2.9
Finance, insurance, and real estate	6.1	5.7	6.0	14.4	8.5	2.4
Finance and insurance	4.1	3.5	4.7	27.3	22.7	5.9
Banks (including Federal Reserve)	1.7	1.9	3.9	26.3	22.4	6.7
Credit agencies	3.1	3.1	3.3	25.0	21.7	7.6
Securities brokers	2.6	3.9	8.3	31.6	23.3	3.8
Insurance carriers	6.0	4.4	7.2	38.0	30.8	5.3
Insurance agents	17.9	15.6	12.0	32.7	20.8	2.7
Investment holding companies	14.2	8.0	10.2	31.2	21.0	3.1
Real estate	6.4	6.1	6.3	9.7	3.4	1.5
Services	5.8	6.5	8.3	16.1	7.9	2.0
Hotels and lodging	0.1	0.1	0.1	1.7	1.6	11.5
Personal	2.7	9.3	13.6	11.5	−2.1	0.8
Business	5.1	7.9	10.9	28.4	17.5	2.6
Auto repair	0.6	0.5	0.2	1.5	1.3	7.6
Miscellaneous repair	0.5	0.4	0.5	2.7	2.2	5.3
Motion pictures	8.3	15.0	31.5	42.2	10.6	1.3
Amusement and recreation	5.3	9.7	12.3	19.7	7.4	1.6
Health	21.2	16.0	19.2	29.5	10.3	1.5
Legal	1.9	2.9	4.0	13.3	9.3	3.3
Educational	10.0	9.5	12.2	46.8	34.6	3.8
Other	19.0	10.0	6.6	10.5	3.9	1.6

[a]1985/1970s average.

NOTE: Figures are averages over designated intervals and are Morgan Stanley estimates derived from the Industry-Commodity Capital Stock Matrix of the U.S. Department of Commerce.

commodity detail is available—the services-providing sector as a whole owned 84 percent of the U.S. economy's total stock of information technology capital. Within services, about 45 percent of this capital can be found in the communications industry (Table 3), largely reflecting the sizable investment in a nationwide telephone system. The finance and real estate sectors are on the next rung down the ladder, accounting in 1985 for about 25 percent of total services sector technology capital. Other large owners of such capital include wholesale trade, business services, and health care providers.

To be sure, the dispersion of information technology capital across the various industries is only a helpful starting point. The key questions that need to be addressed in assessing the role of these technology products really have more to do with how industry has changed its reliance on such items as a factor of production. Table 5 deals with this issue by providing a detailed analysis of the share of each industry's (or industry grouping's) total capital stock that can be accounted for by information technology products, measuring what is called "technology endowment."

Table 6 summarizes this analysis in a systematic and consistent fashion by estimating what can be referred to as "technology intensity measures" (TIMs). Two different TIMs have been constructed in an effort to provide a uniform assessment of relative shifts in information technology endowment. TIM1 is each industry's change in technology endowment—the 1985 level relative to that which prevailed, on average, over the 1970s—compared with the typical change in the economy at large; it essentially measures the relative speed by which various segments have increased the technology share of their capital stock. TIM2 is each industry's 1985 technology endowment relative to the economy-wide average. Thus, each of the TIM variants captures a somewhat different dimension of the shift into information technology. TIM1 looks at *movements* in the technology content of each industry's capital stock, whereas TIM2 is more of a *static* picture of the relative dispersion of technology endowment in 1985.

Communications

As Table 5 indicates, no services-producing segment can match the technology dependency of the communications sector. Over 50 percent of the total capital stock owned by the telephone and telegraph industry can be classified as information technology items—a share that has risen steadily over the past 35 years. This sector's leading role as a high-technology user reflects, of course, the telephone industry's enormous stock of communications equipment that dates back well before the advent of modern-day telecommunications and computers. Thus, this is a mature industry as seen from the standpoint of shifting technology dependence. Because of this characteristic, TIM1—our barometer of relative changes in technology inten-

TABLE 6 Technology Intensity Measures (TIMs)

Business Sector	TIM1	TIM2
ALL INDUSTRIES	1.0	1.0
Goods producing	1.4	0.5
Services providing	0.9	1.2
Transportation	0.9	0.1
Rail	0.4	0.0
Nonrail	1.3	0.1
Air	3.4	0.2
Trucking	1.7	0.0
Other	0.8	0.1
Communications	0.6	4.3
Telephone and telegraph	0.6	4.4
Broadcasting	0.8	2.6
Public utilities	1.3	0.3
Electric	1.2	0.3
Gas and other	2.3	0.2
Total trade	2.1	0.9
Wholesale trade	2.1	1.8
Retail trade	1.3	0.3
Finance, insurance, and real estate	1.1	1.2
Finance and insurance	2.7	2.2
Banks (including Federal Reserve)	3.1	2.1
Credit agencies	3.5	2.0
Securities brokers	1.8	2.5
Insurance carriers	2.4	3.1
Insurance agents	1.3	2.6
Investment holding companies	1.4	2.5
Real estate	0.7	0.8
Services	0.9	1.3
Hotels and lodging	5.3	0.1
Personal	0.4	0.9
Business	1.2	2.3
Auto repair	3.5	0.1
Miscellaneous repair	2.4	0.2
Motion pictures	0.6	3.4
Amusement and recreation	0.7	1.6
Health	0.7	2.4
Legal	1.5	1.1
Educational	1.8	3.8
Other	0.7	0.8

NOTE: TIM1 is each industry's change in technology endowment from the 1970s to 1985, relative to the average change for all industries. TIM2 is each industry's 1985 technology endowment relative to the all-industries average. Technology intensity measures are Morgan Stanley estimates derived from the Industry-Commodity Capital Stock Matrix of the U.S. Department of Commerce.

sity—ranks communications at the bottom of the services sector in terms of its ability to increase its reliance on high technology as a factor of production. Quite simply, since the stock of information technology capital is so large in communications, it cannot possibly grow, for all practical purposes, at the rate experienced by other services providers.

Finance

As TIM2 shows, the finance sector is second only to communications in the degree of its present dependence on information technology. As Table 5 indicates, banks, credit agencies, security brokers, and investment holding companies currently have about 25–30 percent of their capital stock invested in such technologies; for the insurance industry the share is even higher. Also, as the results of TIM1 confirm, this sector has made by far the most rapid move into information technology over the present decade; by industry, credit agencies and banks have increased their technology endowment most dramatically, followed, in order, by insurance carriers, security brokers, and investment holding companies. These results demonstrate quite clearly that of all the services-providing industries, the finance sector has wagered the largest bet on the ultimate productivity payback from information technology.

Trade

As Table 5 shows, wholesale trade establishments have also experienced a sizable increase in their reliance on information technology as a factor of production; this trend stands in sharp contrast to the retail sector, whose dependence on such capital has increased only modestly in recent years, despite the growing profusion of automated end-sales and inventory control equipment. A chronic problem with trade sector data, however, has been an inability to make accurate sampling distinctions between wholesalers and retailers; consequently, for the purposes of this analysis, it seems best to treat that segment of the economy as a unified industry division. In this context, while TIM1 suggests that the trade sector has moved vigorously to acquire information technologies, TIM2 indicates that this surge has come off a low base; indeed, compared with other industries, the trade sector's endowment of such technology still appears a bit on the low side.

Miscellaneous Services

A final grouping of miscellaneous services providers accounts for the remainder of the services sector's spending on information technology. As Table 3 indicates, from the standpoint of aggregate technology outlays the only areas of significance in this collection of industries are business services and health care. The TIM1 measure for the former indicates that this industry has experienced a relatively rapid shift into technology; as TIM2 shows, the speed of this move has been sufficient to leave business services with one of the services sector's highest relative endowments of spending on information technologies. In health care, the shift into such technology has been less dramatic, but that is probably because this industry has always had a fairly hefty endowment of such capital as seen from a TIM2 perspective. Technology ownership in other services-providing industries—whether transportation, public utilities, law firms, or educational institutions—constitutes a very small slice of the U.S. information technology pie and thus is not a real factor in the ''high-teching'' of the services sector. Slight exceptions can be made for the legal and educational groupings where TIM1 findings point to very sharp increases in technology intensity during the present decade.

THE COMPUTER REVOLUTION

Not all technology is alike, of course. The computer is widely perceived to offer the greatest opportunities for improved efficiencies in the workplace, and, not surprisingly, computers are the single largest line item in the information technology budgets of services-providing industries.[1] Excluding the communications sector and its nationwide telephone system, computers total 49 percent of all services sector information-producing capital. That ratio has risen significantly over the present decade and compares with an average of only 17 percent during the entire 1970s. In fact, over the past 15 years, growth in the stock of computers for services has averaged 22 percent per year, almost three times the gains for other technology items and more than five times the growth pace of all services sector capital.

By industry, the computer story actually differs little from that suggested by our previous analysis of total services sector information technology. The finance sector, which owns 49 percent of all services sector computers, also has the highest dependence on this item as a factor of production. Financial services, as of 1985, are estimated to have 27.3 percent of their total capital held in the form of computers and other office machinery—a share over seven times the ratio for all industries. Within the finance sector, securities brokers, investment holding companies, and insurance companies have the highest dependence on computers, followed in turn by banks and credit agencies.

Outside of finance, wholesale trade and business services appear to be most reliant on computers. However, in both of these instances, the computer share of total capital is about half the size of that in the finance sector. Under miscellaneous services, computer ownership is generally small, although both legal services and educational institutions have accelerated their purchases greatly over the present decade. The only exception is communications where, in 1985, more than 98 percent of technology stock was held in the form of telecommunications equipment. Thus, the communications industry, with its relatively small endowment of computers, is not a major factor when the analysis is framed exclusively in terms of the computer portion of the total stock of information technology.

THE PRODUCTIVITY PUZZLE

There can be little debate over the services sector's mounting reliance on information technology. The big question of course is, What does the U.S. economy have to show for these efforts? The answer is not much—at least, not yet.

Table 7 matches up TIMs' estimates against the available industry record on services sector productivity performance. The results are uniformly disappointing. Services providers as a whole, with an information technology endowment that is at present 2½ times the size of that in the goods sector (as shown by TIM2), have experienced a clear slowdown in productivity

TABLE 7   Technology and Productivity Disappointments in the Services Sector

Business Sector	Technology Intensity Measures		Average Productivity Growth (%)		Change[a] (percentage points)
	TIM1	TIM2	1973–1979	1979–1985	
ALL INDUSTRIES	1.0	1.0	0.6	1.1	0.5
Goods producing	1.4	0.5	0.5	2.2	1.7
Services providing	0.9	1.2	0.7	0.4	−0.3
Transportation	0.9	0.1	1.5	−1.2	−2.7
Rail	0.4	0.0	1.1	3.5	2.4
Nonrail	1.3	0.1	1.7	−1.8	−3.5
Communications	0.6	4.3	4.3	3.9	−0.4
Public utilities	1.3	0.3	0.3	1.8	1.5
Trade	2.1	0.9	0.8	1.3	0.5
Finance and insurance	2.7	2.2	−0.1	−1.3	−1.2
Real estate	0.7	0.8	−0.2	−1.4	−1.2
Miscellaneous services	0.9	1.3	0.2	0.7	0.5

[a]Difference in percentage points between 1973–1979 period and 1979–1985 period.
NOTE: TIM1 is each industry's change in technology endowment from the 1970s to 1985, relative to the average change for all industries. TIM2 is each industry's 1985 technology endowment relative to the all-industries average. Technology intensity measures are Morgan Stanley estimates derived from the Industry-Commodity Capital Stock Matrix of the U.S. Department of Commerce; productivity detail is taken from Multiple Productivity Indexes, published by the American Productivity Center based on U.S. government statistics.

growth so far this decade. Moreover, that deterioration follows a period of relatively meager productivity increases over most of the 1970s. The technology "leader"—the finance sector—is estimated to have experienced an especially dramatic deterioration in productivity performance in recent years. In addition, whereas productivity growth is estimated to have picked up slightly in trade and miscellaneous services, after barely rising at all during the 1970s, such improvement seems paltry when compared with the rapidly growing intensity of technology endowment in each of these segments of the economy.

Of course, there are those who say in response that the problem is one of measurement. After all, can anyone really calculate services sector productivity? The difficulty with such a claim is that measurement problems have always plagued estimates of services sector output. The question is: Have measurement pitfalls worsened appreciably during the past decade—a period when productivity growth has deteriorated and technology spending has accelerated most dramatically? There is no conclusive evidence to suggest that this is the case. Most likely, measurement errors have cumulated over a long period of time and, thus, have had a much greater impact on the level of productivity than on its rate of change over the past several years. Thus, it seems an extremely remote possibility that the services sector's productivity disappointments of the 1980s would disappear if the numbers could, in fact, be corrected for statistical problems that date back well into the 1950s.[2]

It is one thing to identify a productivity problem, but it is another task altogether to explain why it occurs.[3] Where the above analysis has been substantiated largely by empirical evidence, efforts to identify the reasons behind services sector productivity disappointments will have to rely mainly on circumstantial evidence. As a consequence, one must conclude with as many questions as answers.

The place to start is with the work force, the human input to any productive endeavor. The services sector's work force is unlike that of any other segment of the U.S. economy. About two-thirds of all services sector employees can be classified as working in the so-called white-collar occupations. In fact, over 80 percent of all white-collar workers in the United States are employed by services-providing industries. By contrast, goods producers have only about one-third of their work force employed in white-collar job categories.

The unique occupational characteristics of white-collar workers underscore the important role played by information technology as a factor in the services sector's productivity puzzle. Services sector "production" turns out to be a process that largely takes place in the office, and white-collar workers— using a rapidly growing stock of information technology—have simply not been able to create the efficiencies that would trigger a meaningful growth in productivity.[4]

There is far more to this problem than just pointing a finger at computers and other information-producing technologies. Admittedly, one of the imponderables simply may be time. As American managers and other "information workers" become more adept at understanding and utilizing complex new technologies, they climb the traditional "learning curve" that eventually opens the door to ever-greater efficiencies. The lack of an instantaneous payback does not necessarily preclude a transitional break-in period that allows information technology to grow naturally into its full productive potential.[5] Another facet of the productivity problem may also be the unique nature of the technology itself. The computer revolution is, in fact, the first major technological breakthrough in the U.S. economy that is being applied primarily to the generation of that seemingly amorphous commodity called services sector "output." By contrast, in the past, other technological breakthroughs have been applied mainly to the production or delivery of goods. Evaluating such technology investment must, therefore, be done in the absence of well-established criteria and guidelines. Consequently, without the historical benefit of such experience, estimates of productivity paybacks from information technology are in many instances a "shot in the dark."

This puts the problem squarely in the lap of U.S. managers, and in that regard the task is a formidable one. A first imperative is to develop an accurate system of measuring white-collar productivity. Only then can the cost of information technologies be evaluated more precisely against the work effort of those employees who rely on such equipment. One of the clear problems in "costing-out" information technology is the failure to recognize that a broad spectrum of workers is involved in its productive applications, ranging from the so-called knowledge workers who control the analytical applications to the vast cadre of management information system support staffs that are required to keep the machinery in operation. Thus, a revision of traditional accounting standards may be an important prerequisite to developing a white-collar productivity measurement system.[6]

Additionally, managers need to give more thought to a strategy of technology acquisition. Fifteen years into a phenomenal technology-buying binge, corporate America appears hooked on each and every twist of the product cycle. The latest in computer gimmickry does not necessarily guarantee obsolescence of an existing installation of information technology. To avoid the temptation of open-ended technology budgets, the replacement cycle needs to be evaluated against more explicit productivity criteria of the user community.

Another area of concern pertains to the "hows" and "whys" of the application of information technology to productive endeavors in the services sector. Two issues come to mind in this regard: product selection and software. From the standpoint of product selection the issue is really that of the general-purpose machine versus the more focused piece of equipment. Technology acquisition needs to be directed more explicitly to the task at hand.

The automated teller machine of the banking industry is an example of a product-specific piece of technology equipment that has a highly focused application; mainframes and other general-purpose computers clearly would be redundant in such a capacity, which is a general lesson that could be relevant for other services sector industries as well.

Software, needless to say, has always been critical to the productivity payback of information technology. To date, progress has been far more rapid in hardware than in programming and operating systems that link information technology explicitly to production. The manufacturing sector has thus far reaped most of the productivity benefits from software break-throughs—especially in areas such as computer-aided design and manufacture, expert systems, and artificial intelligence. By contrast, the services sector clearly has been laggard in exploiting the efficiencies of technology applications.

This discussion has been aimed at developing a better understanding of technology's disappointing productivity payback in the services sector. Yet given the diversity of industries that provide services in the United States, it is quite possible that information technology just is not the answer in some areas. For transportation and lodging, it is obvious that greater efficiencies in computer-based reservation systems offer a real opportunity for greater booking volume per unit of worker input. An analogous conclusion can be drawn for the role that information technology plays in facilitating a greater volume of transactions in the financial and trade sectors.

What about lawyers, consultants, investment counselors, educational institutions, and other professionals who produce information as a final product? Is it really in the economy's best interest to equip these services providers with efficient technologies that have the potential for overwhelming the economy with information? Spreadsheets and word processing are remarkable tools, but if they are abused, the economy could find itself engulfed in an ocean of information. The potential is "information overload," and questions in this regard obviously go well beyond the productivity challenges facing individual industries. It may well be that many services sector activities should also be scrutinized from the "macro" standpoint of how they affect the productivity of those who purchase such services—an admittedly formidable task.

Whatever the answer for the services sector, it seems increasingly clear that the high-technology revolution of the information age has yet to deliver from the standpoint of national productivity enhancement. In this regard, it also seems increasingly clear that the link between technology and productivity performance is central to the present debate on America's competitive dilemma. The lingering productivity shortfall in the United States is no longer a problem traceable mainly to manufacturing, since that sector has now moved back to its longer term productivity growth trend. However, the services sector, which currently accounts for about two-thirds of private sector output and employment in the United States, has not been so fortunate, and it is safe to say that this now dominant segment of the economy has to bear greater responsibility for lagging national productivity growth. Unfortunately, for the economy as a whole, there may be small consolation in the competitive renewal taking place in manufacturing. If services fail to carry their weight, renewed factory sector vitality simply will not be enough to return the U.S. economy to its historical growth path. This is the essence of the United States' hidden competitive challenge.

NOTES

1. Conversations with government officials indicate that the "other office machinery" category accounts for a very small portion of the total, in terms of both the stock and its rate of growth.
2. A similar conclusion can be supported by the analysis of Mark in his chaper (this volume). This analysis does suggest that measurement problems have worsened in commercial banking in recent years; however, this industry accounted for only 11.4 percent of total finance sector output in 1986, hardly enough to alter the dramatic productivity disappointments for this segment as a whole.
3. This problem has plagued the productivity literature for years. See, for example, Baily (1981), Denison (1979), and Wolff (1985).
4. Estimates show that white-collar productivity has declined steadily over most of this decade to levels last seen in the late 1960s; this result is developed more fully in Morgan Stanley (1987).
5. Elsewhere it has been argued that today's computer revolution is comparable to the industrial revolution, the building of a nationwide rail system, and the advent of the factory assembly line—all dramatic technological breakthroughs that sparked a process of structural change that paid off only after a considerable period of time; see Morgan Stanley (1986).
6. For an important start in this direction, see Strassman (1985).

REFERENCES

American Productivity Center. 1950–1985. Multiple Productivity Indexes.

Baily, M. N. 1981. Productivity and the services of capital and labor. Brookings Papers on Economic Activities (1):1–65.

Bowen, W. 1986. The puny payoff from office computers. Fortune 113(May 26):20.

Denison, E. 1979. Explanations of declining productivity and growth. Survey of Current Business (August):1–24.

Morgan Stanley. 1986. The productivity puzzle: Perils and hopes. Economic Perspectives (April 10).

Morgan Stanley. 1987. America's technology dilemma: A profile of the information economy. Special Economic Study (April 22).

Musgrave, J. C. 1986. Fixed reproducible tangible wealth in the United States: Revised estimates. Series in Survey of Current Business (January):51–75. Washington, D.C.: U.S. Department of Commerce.

Neikirk, W. R. 1987. Technology in America, a five-part series. Chicago Tribune (July 5–12).

Schneider, K. 1987. Services hurt by technology. *The New York Times* (June 29):23.

Strassman, P. 1985. Information Payoff: The Transformation of Work in the Electronics Age. New York: Free Press.

Wolff, E. 1985. The magnitude and causes of the recent productivity slowdown in the United States: A survey of recent studies. Pp. 29–57 in Productivity Growth and U.S. Competitiveness, W. J. Baumol and K. McLennan, eds. New York: Oxford University Press.

Measuring Productivity in Services Industries

JEROME A. MARK

The increased importance of services industries over the last two decades and concern over U.S. productivity growth have stimulated interest in productivity measures for this sector of the economy. As a result, the Bureau of Labor Statistics (BLS), which has responsibility for developing the government's measures of productivity, has concentrated a great deal of its efforts on expanding the number of services industries for which it publishes productivity data. These data are in the form of indices of output per unit of input derived from dividing an index of output for an industry by the corresponding index of input.

This has been and is a difficult undertaking, but the number of services industries for which productivity measures are available has increased substantially in recent years. Nevertheless, much remains to be done. This chapter describes these efforts, including some of the problems of measuring productivity in services industries and the way in which some of the measures have been developed. It also presents the results of BLS work, including the movements of productivity in the various services industries.[1]

Some economists have argued that the productivity slowdown in the United States over the last decade and a half may have been more apparent than real because of the increased importance of the services sector and the weaknesses of the productivity measures in services industries. Others have questioned whether productivity measures can be derived at all for services industries because of the very nature of their activities. The traditional arguments have generally been that the outputs of goods industries are tangible and storable, and therefore measurable, while those of services industries are neither, and

139

therefore not measurable. Related to this is the belief that units of services are less homogeneous than units of goods, which reflects the greater differences in quality among units and thereby presents additional problems of measurement.

There are indeed serious problems of measurement for some parts of the services sector, and it may not be possible even with intensive study to resolve some of the conceptual difficulties or to develop the data necessary for reliable measures of output and, in turn, productivity for these activities. The lack of homogeneity in many legal, medical, educational, and entertainment services, for example, clearly presents difficulties of measurement that may preclude the derivation of satisfactory productivity measures.

On the other hand, there are many parts of the services sector for which the problems of measurement are no more severe than for parts of the goods-producing sector. Successful approaches for developing measures of goods activities can be applied to many services activities.

To understand the diverse movements of productivity among industries, it is important to attempt to develop separate productivity measures for services industries and, to the extent possible, resolve the difficulties of measurement. That is what the BLS has been attempting to do.

DEFINITIONS

In this chapter the services sector is defined broadly to include all non-commodity-producing industries. This encompasses the major industrial groupings of transportation, communications, electricity, gas, and sanitary services (public utilities), trade, finance, insurance, real estate, and government, as well as business and personal services. The industries excluded from this group are those in agriculture, mining, construction, and manufacturing. Other definitions have been used by private groups presently included in this definition. For example, Fuchs (1968) in his work on services industries' output and productivity measurement excludes transportation, communications, and public utilities. Marimont (1969) in describing output measurement in services industries in the national income and product accounts limits his coverage to finance, insurance, real estate, and business and personal services. Any definition is perhaps arbitrary. The BLS also uses the broader definition to ensure inclusion of as many industries as possible.

Productivity measures are those that relate physical output to physical input. As such, they encompass a family of measures including single-factor input productivity measures, such as output per unit of labor or output per unit of capital input, as well as multifactor productivity measures, such as output per unit of labor and capital combined.

The most extensively developed and widely used productivity measure is the one relating output to labor input. It is a measure relevant for analyzing labor costs, real income, and employment. It is also the type of productivity indicator for which data are more readily available to derive adequate measures. As a result, almost all BLS services industries productivity measures developed are of this type and the ones on which this chapter concentrates. Work has been undertaken to develop multifactor productivity measures including capital, materials, and purchased services inputs, as well as labor for telephone communications and for electric, gas, and sanitary services, but these measures are not yet available.

Changes in output per employee hour, as with all single-factor productivity measures, do not imply that labor is solely responsible for the changes in productivity. Movements in output per hour reflect technological innovations, changes in capital input, scale economies, education, management, and other factors, as well as skills and efforts of the work force.

PROBLEMS OF MEASUREMENT

In many ways, the problems of measuring output and, in turn, productivity for services industries are similar to those for goods industries. That is, the output indicator must be quantifiable and independent of the input measure. An industry output measure that is based on an input measure, as it is in some instances in the national income and product accounts (e.g., the products of general government, households, and nonprofit institutions) obviously results in an incorrect measurement of change in productivity for that activity. Similarly, the coverage of the output measure should be the same as that of the labor input measure. If not, an imputation is implicitly being made that the movement of the portion of the input that is not included in the coverage of the output measure is the same as the covered portion.

It is important also to distinguish between intermediate and final services. Productivity measurement attempts to ensure that the indicators used represent output flowing from the industry being measured and are not part of an intermediate step in the services flow. In this sense, productivity measurement differs from work measurement, which generally refers to the analysis of the labor requirements at operational stages of an activity. For work measurement the technology associated with the activity is fixed. Productivity measurement refers to the final services provided by the organization or industry and their relationship to input. Changes in these measures do reflect, among other things, changes in technology.

OUTPUT

In the case of an industry providing one type of service, output is merely a count of the units of this service, however defined. This assumes that there is homogeneity in the service being counted. In the more usual case of an industry providing a number of heterogeneous services, the various units must be expressed in some common basis for aggregation. In measures of an output per unit of labor input, this basis is in terms of the base year labor input requirements for the different types of services. In this way, the output measures for developing labor productivity measures differ from the more traditional production measures which are based on total price or value-added weighting of the components.[2]

When there are quality changes within a service, adjustments must be made in the output measure to account for the fact that the service is no longer the same homogeneous unit. However, the meaning of quality change for labor productivity measurement differs from the usual concept of quality change associated with consumer price measurement in that it reflects differences in producers' labor requirements or labor costs rather than differences in consumer utility.

This difference in the meaning of quality change for productivity measurement versus consumer price measurement is illustrated in the goods area with regard to the introduction of the catalytic converter in automobiles. The catalytic converter was added to automobiles in the 1970s as a device for meeting the antipollution requirements mandated by the government. The cost of purchasing an automobile increased when the device was added. The utility or satisfaction to the consumer who purchased the automobile did not increase, and in developing a measure of changes in consumer prices as an indicator of changes in the cost of living, this could be viewed as a price increase (similar to a tax). From the point of view of the producer, this change represented a quality increase. Labor and capital resources went into the production of the item, and it was an addition to the car—in effect there was now more car per car. After careful consideration the BLS, which is responsible for providing both the government's consumer price indices and productivity indexes, decided to treat it as a quality increase.

Ideally, then, output measures should incorporate data on the number of services provided differentiated by unit labor requirements and in sufficient detail to account for quality differentials. In practice, however, such data are generally not available. As a result, approximations based on alternative approaches utilizing various assumptions are used.

In the absence of quantitative information on the units or amount of services, the principal alternative is to remove the change in price from the change in value (reflecting both price and quantity) of the volume of services. This approach is tantamount to weighting the quantities of services provided with price weights. Insofar as price relationships among the various component services of a services industry are similar to unit labor requirements or unit labor costs relationships, this measure approximates the desired measure. Also, since it is generally easier to measure price change for services defined with detailed specifications, this approach is most generally used when adequate quantity information is not available.

However, this approach requires price data in sufficient detail to represent adequately the price trends of services included in the change in value of the services. Otherwise, price movements of the covered areas are implicity imputed to the uncovered areas. However, because the relationship among the price movements of similar services is stronger than the relationship among quantity changes of various services, this alternative has greater viability than imputing quantity changes of covered areas to those of uncovered areas.

Nevertheless, the use of price deflators still requires ideally that adjustments for quality change be made. As mentioned earlier with regard to direct quantity measurement, this adjustment to the price measure should also be made on a cost basis.[3]

In actual practice, the BLS output measures for its services industries productivity measures are mixtures. Some, such as those for transportation and public utilities industries, are based on quantity data. Others are based on price deflation because of inadequate quantity information. Others utilize the deflation approach at lower levels of aggregation and labor input weighting at higher levels of aggregation.

LABOR INPUT

To derive a productivity measure that relates the output to the corresponding labor involved in the production or services-generating process, it is important to have data on the hours worked by all persons involved in the production process. In addition, the hours should refer to hours worked differentiated by types of employees in the particular industries. Unfortunately, the data available have serious gaps for meeting these requirements.

The data required cover the hours of nonsupervisory, supervisory, and unpaid family workers, as well as the self-employed. The principal source of data on employment and hours is the BLS survey of establishments' payrolls, the Current Employment Survey (CES). This survey provides good measures of employment and hours of nonsupervisory workers by industry, but it does not provide data on the average hours of supervisory workers or the employment and hours of the self-employed and unpaid family workers. Information on the self-employed and unpaid is derived from a survey of households of the noninstitutional population, the Current Population Survey (CPS). These data, which are based on a survey of 60,000 households in the United States, are adequate for the measures of the business economy and major sectors but present limitations when used for measures of services industries.

At the present time, the average weekly hours of supervisory workers are assumed to be equal to those of nonsupervisory workers in services industries. This assumption presents fewer limitations for developing measures of change than for developing measures of levels.

As mentioned earlier, a desirable measure of productivity is one that reflects the change in labor input actually involved in the generation of the services provided. The hours data in the CES are based on hours paid and include paid vacations, holidays, sick leave, and other time off in addition to actual hours worked. To the extent that leave practices change, the resultant productivity measures over- or understate the actual change in output per hour.

To develop a better series of hours at work, the BLS has been conducting an annual survey since 1982 of some 4,000 establishments to collect data on hours at work and hours paid for nonsupervisory workers in the private nonagricultural business sector. From this survey, ratios are developed to adjust the present hours-paid measures to an hours-at-work basis. The definition of hours at work was established, after careful study, as time on the job or at the place of work. It includes coffee breaks, short rest periods, paid cleanup time, and other paid time at the workplace, in addition to actual time worked. This definition was considered to be conceptually the most acceptable one for which data could be extracted from establishment records.

Although the appropriate hours information is currently available from this survey for the aggregate measures, a substantial expansion of this survey will be required to develop reliable data for specific services industries. This will be very costly.

The BLS measures of productivity based on the hours of all persons assume that workers are homogeneous with respect to skill. However, a highly skilled worker can be viewed as providing more labor services per hour than a less skilled one. When skill differences are ignored, increases in skill levels are measured as increases in productivity. As a result, shifts from less skilled to more skilled labor because of increased education or experience are not reflected as increases in the measures of labor input.

This would not be a problem if the proportions of workers at different levels of productivity were constant over time. However, to the extent that there are changes in the composition of the work force with respect to education and experience, which result in skill differences, it may be desirable to adjust the labor input measure for these changes, which otherwise would be reflected in the productivity measure.

To address this problem, previous studies have usually taken the position that relative wage or income level differentials associated with specific worker characteristics reflect marginal productivity of these attributes. Generally, the included characteristics are the number of years of schooling, age, sex, and possibly industry and occupation (Gollop and Jorgenson, 1980). Weighting the quantity of labor (measured in hours or employees), classified by these characteristics of the work force, by relative wage or income differentials results in an aggregate measure of labor input intended to reflect the composition of the work force.

This procedure is not without problems. For example, workers with similar characteristics have different earnings in different occupations and industries. However, this correlation between industry or occupation and earnings may also be due to influences other than productivity, such as differences in the cost of living or degree of unionization.

The BLS is currently developing new measures of labor input based solely on changes in the amount of work experience and schooling workers acquire (Waldorf et al., 1986). The methodology used follows directly from the economic theory of human capital developed by Mincer (1974) and Becker (1975). It assumes that increased schooling and on-the-job training increase one's stock of skills and productivity. It also assumes that economic returns to higher education and additional work experience reflect the marginal productivity of these characteristics. The BLS has developed a multidimensional data base which cross-classifies the annual hours of workers grouped by schooling and experience. The data base is developed from various models that make use of decennial census data, a matched sample from the CPS and from social security records.

As mentioned above, it is recognized that hourly wages differ not only because of skill differences but also because of factors unrelated to productivity. Accordingly, simple averages of hourly wage rates for each education and experience group are not necessarily suitable approximations of marginal productivity. Consequently, the BLS has developed an econometric model that provides measures of wages dependent on changes in education and experience but simultaneously controls for other types of variation.

Skill-adjusted labor input measures have been developed for the business sector, and this work is currently being extended to determine the feasibility of developing corresponding measures for specific industries, especially services industries.

MEASURES OF PRODUCTIVITY FOR SERVICES INDUSTRIES

At present, the BLS publishes indices of output per hour (i.e., output per unit of labor input) for industries in each of the major services activities. This includes all groups of Standard Industrial Classification (SIC) 40 and higher: transportation, communications, public utilities, trade, finance, insurance, real estate, and business and personal services. Productivity measures for the federal government are also published with separate detail for the common functions provided by federal agencies, such as recordkeeping, library services, buildings and grounds maintenance, and loans and grants.

In addition, work is currently being conducted on the development of indices of productivity for hospitals, wholesale trade activities, and automobile repair shops.

The following sections describe the procedures utilized and some of the problems present in developing the particular measures. Because the methods and the sources used for deriving the labor input measures are the same for each of the separate industries, this discussion focuses on the output measures developed for the productivity indices.

Transportation

The BLS publishes productivity indices for five transportation industries: railroads, intercity trucking, intercity buses, air transportation, and petroleum pipelines. These measures cover 57 percent of transportation employment.

Conceptually, the measures for transportation industries are easier to develop than those for other services industries because the transportation industry output—the movement of goods or passengers over distances—is more easily quantified. Output units in transportation have two dimensions, amount and distance, that reflect how much has been transported how far. As such, ton-miles, passenger-miles, barrel-miles, and so forth are the primary output indicators for these services.

Historically, these data have been available mostly from the regulatory agencies of the transportation industries. In many cases, however, there are data gaps that place certain limitations on the measures. For example, it is sometimes impossible to adjust the output measures adequately for changes in the average length of haul. The unit labor requirements associated with the movement of freight and passengers are usually greater for short hauls than for long hauls. Therefore, a shift from a long haul to a series of short-haul trips or vice versa could be interpreted as a change in productivity, although only the mix of trips had changed.

For the two major freight-carrying industries, railroads and trucking, undifferentiated ton-mile data are reported for total freight operations. In trucking, the ton-mile data are also reported separately for general, contract, and other carriers. However, the output measures should reflect the kinds of commodities handled and the average distances they are moved since these represent separate types of services. The preferred way to develop such measures would be to combine the tonnage and average haul of each commodity by its respective labor input requirements and aggregate the results for all commodities transported. Unfortunately, this cannot be done with available data.

However, separate commodity information on tonnage for railroads is available from the Interstate Commerce Commission for about 200 commodity lines, ranging from agricultural and mining products to motor vehicles and scientific instruments. Several years ago similar information was available for the trucking industry, but it was discontinued. The BLS uses these data to adjust the overall measure of freight ton-miles for changes in the composition of goods carried.

Although this commodity adjustment represents an improvement over undifferentiated ton-miles, refinements cannot be developed to the extent desired. The commodity index adjustments are made in terms of unit revenue weights. The underlying assumption, therefore, is that differences in labor requirements among commodities are similar to differences in unit revenues. Since labor costs constitute more than half of each industry's total operating costs, this assumption may not be unreasonable. However, the proportions could conceivably differ by commodity.

In deriving the total industry output index for each of the transportation industries, the freight ton-miles measure (adjusted or not) is combined with the revenue passenger-miles measure.

The deregulation of many transportation industries has resulted in elimination of some of the operating statistics that were previously published and used to develop output indices. As a result, some of the transportation industry measures have had to be extended on the basis of more limited information. The BLS has been cooperating with other government agencies to ensure that adequate statistics for transportation industries remain available.

Communications

The BLS productivity index for telephone communications covers about four-fifths of the employment of the communications sector. The output index is derived from revenues of all telephone companies reporting to the Federal Communications Commission. The revenues are stratified by major source (i.e., local, toll, and miscellaneous) and deflated by specially prepared price indices reflecting these different services.

The deflated revenue measure includes revenues from private line services. It also accounts for television, radio, and computer transmission by telephone industry facilities, as well as directory services.

Despite the details included in the output measure, improvements could be made if information were available on the intensity of use of telephone equipment by customers. The number of calls made can vary without revenue varying proportionately because of flat charges, such as WATS (Wide-Area Telecommunications Service) line or local call charges. To the extent that the number of such calls varies over time, the index over- or understates output change.

Electric, Gas, and Sanitary Services

Services rendered by public utilities range from the provision of light, heat, and water to the disposal of solid and liquid wastes. In this area, the BLS currently publishes productivity indices for electric utilities, gas utilities, and a combination of the two.

The measure of electric utility output covers all privately owned utilities, which account for roughly four-fifths of the total output generated in the United States. Output is defined in terms of kilowatt hours generated and distributed. The measure of output of gas utilities is defined in terms of therms (1 therm equals 100,000 British thermal units) delivered to customers by all privately owned companies (which account for 95 percent of total gas output).

Since the labor requirements per kilowatt hour or per therm are higher for residential than for commercial and industrial customers, and higher for small establishments than for larger ones, the BLS differentiates both the kilowatt-hours and the therms by type of customer.

The BLS is currently attempting to develop a productivity index for sanitary services, including sewage and refuse systems. For this measure, output is measured in terms of disposal of metered liquid sewage and tons of solid waste processed differentiated by type of waste insofar as different types of solid waste require more or less labor per unit processed.

Retail Trade

The BLS has been publishing measures for retail trade industries since 1975; however, the number of industries has increased markedly in recent years. At present, indices are available for 12 important industries: retail food stores, franchised new car dealers, gasoline service stations, apparel and accessory stores, furniture and appliance stores, eating and drinking places, drug and proprietary stores, and liquor stores. Apparel and accessory stores are further broken down into men's and boys' clothing stores, women's ready-to-wear stores, family clothing stores, and shoe stores. Also, furniture and appliance stores are disaggregated into furniture and furnishings stores and appliance stores. Work is currently under way on measures for hardware stores, auto and home supply stores, and department stores.

For most retail trade industries, data on gross sales in current dollars, deflated by appropriate price indices, are used to estimate real output. This method can yield good measures of real output if adequate price indices reflecting the price movements of the various commodities sold by the establishments can be developed. The recent improvements that have been made in the BLS price index program have enabled it to develop output and productivity indices for more components of retail trade.

Productivity indices based on deflated value of sales output measures reflect shifts among services with different values but the same trade labor requirements. Therefore, the overall industry productivity index can show movements without any change in component elements.

In retail industries, a large portion of the value of sales is provided by the manufacturers and the wholesalers of the products sold. A net output measure would be desirable because it would correspond most closely to the functions provided by the retailer. Unfortunately, net output measures based on separately deflated final sales and cost of materials data can result in large errors of measurement when the cost of materials is a large proportion of the final value. All the errors in the current value of sales, the current value of goods purchased, the product price indices, and the materials price measures affect this residual. A gross or total sales measure will yield the results of a net or value-added measure with perhaps less measurement error if the value added as a percent of sales (gross margin) does not change over time. Available data indicate that for most of the industries published, average gross margins have not changed substantially over time. In order to introduce labor input weighting at some stage, the indices for retail trade industries for the most part are developed in two stages. Deflated output measures are first developed for detailed merchandise lines. These are aggregated to higher levels, and the resultant indices are then combined with labor cost weights. For example, in retail food stores, sales for 13 key merchandise lines are deflated by using specially prepared price indices based on the BLS Consumer Price Index components. The merchandise lines indices are aggregated into five department lines: meat, produce, frozen food, dry groceries and dairy, and all others. These, in turn, are aggregated with labor cost weights to develop the overall output measures for grocery stores.

Wholesale Trade

At present, the BLS does not publish any indices for industries in the wholesale trade sector. It is currently examining data for four industries in considerable detail in order to derive reliable measures: metal service centers; scrap and waste materials dealers; petroleum bulk stations; and beer, wine, and distilled spirits distributors. These industries include about half a million workers or 10 percent of the employment in the sector.

Physical quantity data are available to develop output measures for these industries. The quantity data for disaggregate commodities will be combined with fixed period labor input weights reflecting the services provided by the wholesalers to retailers and other users.

Several measurement problems concerning these industries need to be resolved. Some firms perform work on commodities that they distribute to retailers, but this varies substantially among wholesale distributors. Whether the labor input weights can adequately take this into account is questionable. In addition, in some instances a regional wholesaler distributes commodities to local wholesalers. This creates a problem of duplicate counting in the overall industry output measure.

Finance

In finance, the BLS publishes a measure for the commercial banking industry. The output measure for this industry is in terms of the three major services that commercial banks render their customers: deposits, loans, and trust services. While banks also perform non-fund-using services, such as safe deposits and customer payroll accounting, lack of adequate data precludes deriving a measure for them. However, because the proportion of employees engaged in such services is small, the overall measure is little affected by the omission.

There has been a great deal of discussion over the years about the appropriate measure of banking output. Much of the discussion narrows down to whether the appropriate concept is one based on what has been called the liquidity approach or one based on a transactions approach (Gorman, 1969). In the former, banks are viewed as providers of money to hold, and their output is measured in terms of the interest received on the volume of deposits. Such interest received by banks is assumed to be equivalent to income foregone by depositors due to their preference for holding their financial assets in banks—as deposits—rather than investing directly. The interest that the depositors forego represents the value of the banks' services in meeting their depositors' preference for liquidity. This approach can be extended to savings accounts and other time deposit accounts, the assumption being that the foregone net interest is the value of the banks' services.

The other approach is that banks are providing a series of services to their clients which are reflected in the transactions performed. The volume of the banks' output is proportional to the volume of transactions handled. In developing the productivity measure, the BLS adopted the second approach for measuring output.

Thus, the final output of banks represents an array of services provided to bank customers relating to depository, lending, and fiduciary functions of banks. Estimates of the number of transactions in each of these services functions are derived. In some instances, no direct count of transactions is available, so the number of transactions is estimated from data on the total value of transactions and on the results of surveys of average transactions amounts. While these estimates have some limitations, a count of the number of transactions is the measure used to reflect the quantities of services provided.

Deposit activity is measured in terms of the number of checks transacted and the number of time and savings deposits and withdrawals. The data for demand deposit activities are derived ultimately from Federal Reserve counts and official benchmark surveys. For time and savings deposit activity, the output measure is derived from data published by the Federal Deposit Insurance Corporation and the Functional Cost Analysis conducted annually by the Federal Reserve Board.

Lending services provided by banks are also measured in terms of units. As in the case of deposit activity and trust department activity, the BLS does not use banks' financial data to arrive at component output measures. Such use of financial data would be misleading even if appropriate price deflators could be developed; a rise in the constant dollar value of the loans might simply reflect a few large loans having been made and a fall could reflect the repayment of a few large loans even as the number of small loans may have increased.

The BLS measures 10 types of loan output, using data generated by the Federal Reserve Board, the Department of Housing and Urban Development, the Federal Home Loan Bank Board, and others. Included in the loan output measure are residential and commercial mortgage loans, consumer loans, single-payment loans, credit card loans, and commercial loans. The number of loans is usually derived by dividing the average face value of a loan into the total value of all the loans in a given category.

The output measure for the fiduciary or trust department services is derived from the number of accounts stratified by five major categories, including employee benefits trusts, personal trusts, and estates.

The output indices for the three segments are then aggregated with employment weights to derive the overall output index.

Federal Government

In addition to developing private sector measures, for the past 14 years, the BLS has been conducting a program of developing labor productivity indices for all federal government agencies with 200 or more employees. Currently, the measures include 390 organizational units representing 59 agencies which cover 69 percent of the civilian work force.

The BLS calculated productivity, output, and employee years indices for the overall sample and for each of the 390 organizational units. Indices for individual organizations are not published but are returned to the responsible agency for its own use. Organizational data are grouped into 28 functional areas, such as audit, electric power, and personnel, for which the indices are published.

Where possible, the relevant concept of output of a government agency is its final product output, that is, what the given organization produced for use outside the organization. In this work, the output data included in the overall sample are, in fact, final from the perspective of the organization and from the functional groupings in which these organizations are classified.

However, since the outputs of one organization may be consumed wholly or partially by another federal organization in the production of its final outputs, all output indicators will not be final from the perspective of the entire federal government. Therefore, the overall statistics presented in the study do not represent federal government productivity, but rather the average of the productivity changes of the measured organizations included in the sample.

The data base for the output indicators is quite extensive, having expanded fourfold since the program's inception. Currently, output information is included on about 3,000 indicators. By expanding the output detail of the survey participants, the BLS is better able to measure the services-oriented activities of the federal government.

The indicators are diverse and illustrate the many functions and services provided by the government. For the auditing function, typically reported indicators are inspections completed and audit reports prepared. Providing information to the public is a major function and is measured by such things as statistical reports issued, maps produced, and weather forecasts made. Regulatory activities cross a wide spectrum and include food inspections conducted, drug arrests made, applications and licenses issued, drugs approved, and patents approved. In transportation and utilities, the measures are similar to those employed for the private sector and include revenue ton-miles and kilowatt-hours. To the extent possible, data in sufficient detail are sought on these indicators to take account of shifts in the mix of services provided.

State and Local Government

More recently, the BLS has begun to explore the possibility of computing productivity indices for state and local government. There are a number of services areas in which final outputs can be specified, including electric power, drinking water, solid waste collection, and alcoholic beverage sales. Most have private sector counterparts. However, there are many services areas in which a lack of research or agreement exists on correct organizational outputs, including such important areas as education, fire, and police.

There is no comprehensive system for collecting data on state and local government outputs. In a few select services areas, including electric power and mass transit, output statistics are collected as part of a federal agency's program and oversight responsibilities. However, in most services areas, output statistics are not available.

Business and Personal Services

In the area that includes not only business, personal, and repair services, but also education, health and social services, and political organizations, the BLS currently publishes measures of productivity for three activities: hotels and motels, laundry and dry cleaning services, and beauty and barber shops. Measures of productivity for automotive repair shops and for hospitals are also being developed. The employment in the industries for which measures have been or are being developed covers about 23 percent of the total employment in this sector.

Because physical output data are not available for the three industries that are presently published, the output indices for these industries are developed by using a deflated value approach. The techniques are similar to those described above in that the changes in revenues for specific services are divided by appropriate price indices and these output measures are then aggregated with employment weights to derive the industry output index.

As mentioned above, the BLS is currently developing a measure of productivity of hospitals. The industry is defined as including all nonfederal short-stay hospitals (i.e., with patient stays of 30 days or less). Because the underlying concepts and measurement procedures are of particular interest and are currently being developed, they are discussed in some detail.

Hospitals provide services designed to eliminate, retard, or neutralize pathologies. They also provide gynecological, neonatal, and other services. Treatments and related procedures can be regarded as the producer technologies by which those services are rendered. Treatments, of course, are specified to produce desired outcomes. However, since outcomes depend on factors other than, and in addition to, treatments—for example, preadmission health of the patient[4]—there is some question whether the output measure related to hospital labor and other resources used for a productivity measure should be based on outcomes (e.g., see Scott, 1979). The BLS does not adjust for outcomes in deriving its output measure for its productivity index.

Illnesses are classified as diagnoses once they have been identified clinically. Diagnoses tend to be standardized, and each diagnostic category or diagnosis-related group (DRG) implies a complex of treatments or procedures, which in turn indicate certain kinds and amounts of resource utilization.

The concept of DRG was evolved by the Health Systems Management Group at the Yale University School of Organization and Management during the 1970s and has been modified gradually. It is based on the several thousand entries in the *International Classification of Diseases* (1978). To begin with, all diagnosis codes have been condensed into 23 mutually exclusive and exhaustive major diagnostic categories (MDC) (U.S. National Center for Health Statistics, 1985). The MDCs are generally based on diseases that tend to be diagnosed and treated similarly by specialists. Hospital discharges are coded first by their pertinent MDCs. Each MDC is, in turn, partitioned into numerous DRGs. The major partitions are between surgical and nonsurgical procedures. Surgical procedures are divided into procedural categories based on resource utilization. Nonsurgical procedures are similarly classified hierarchically with further partitioning by age, comorbidities (having more than one disease), and complications. Comorbidities and complications are defined in terms of an increase in length of stay of one day or more, over and above the standard length of stay for a given DRG.

The output measure is derived from data on number of inpatient discharges and number of outpatient visits weighted on the basis of revenue weights. The data on inpatient discharges are drawn from the National Discharge Survey of the National Center for Health Statistics of the U.S. Department of Health and Human Services and are classified by diagnostic category.

The discharges in each diagnostic category are aggregated with weights based on the average hospital operating costs per diagnostic category. The weights are derived from data from the Blue Cross Federal Employee Plan for the period 1972–1978 and, after that, from data on average cost relatives for DRGs determined by the Health Care Financing Administration (HCFA) for Medicare and various state commissions. The cost relatives reflect labor costs and exclude capital costs.

Varying intensities of care are captured by the output measure because of the different DRG weights assigned to the various treatment modes. For example, intensity of care varies substantially according to whether or not surgery is performed. The separation of MDCs and, in turn, DRGs into surgical and nonsurgical classifications takes this into account. Rising intensities have been partially offset by declining lengths of stay, which are also reflected in the DRGs.

When a new service is introduced—for example, a new imaging device— intensity of care may not be affected. If it is affected, treatment protocols may not be. When a shift occurs from invasive to noninvasive procedures, such as from surgery to remove kidney stones to extracorporeal shock-wave lithotripsy, intensity declines. A new DRG will be introduced or the weight of an existing one recalibrated. If so, the new weight is introduced into the output measure.

As with all BLS industry productivity measures when they are being developed, the hospital measure is currently being reviewed by industry and labor representatives. These reviews provide insights into problems not anticipated and suggestions for improving the measures that are being developed.

WHAT DO THE DATA SHOW

Table 1 shows the productivity growth rates of the various services industries for which the BLS has published measures. As can be seen, the growth rates of the services industries vary substantially. Some, such as

TABLE 1 Productivity in Services Industries for which BLS Productivity Measures Are Available

Industry	1965–1973	1973–1985	1973–1981	1981–1985
Transportation				
Railroads	4.2	4.6	2.3	10.8
Bus carriers	−1.5	−1.0[a]	−1.4	−0.3[b]
Intercity trucking	2.7	0.4	0.5	0.6
Postal services	1.3[c]	1.3	1.4	1.0
Airlines	5.3	3.9	3.5	6.8
Petroleum pipelines	7.9	0.1	−0.7	5.6
Communications				
Telephone communications	4.7	6.2	6.7	5.1
Public utilities				
Gas and electric utilities	4.9	−0.5	0.3	−0.6
Electric utilities	5.4	0.2	0.8	0.8
Gas utilities	3.9	−2.1	−0.4	−4.4
Trade				
Retail food stores	2.2	−1.0	−1.1	−0.5
Franchised new car dealers	2.6	1.2	0.6	3.3
Gasoline service stations	4.9	3.2	3.2	4.2
Apparel and accessory stores	2.8[c]	3.9	3.0	5.3
Men's and boys' clothing stores	3.3[c]	2.3	1.4	3.4
Women's ready-to-wear stores	4.6[c]	6.0	4.5	6.4
Family clothing stores	5.9[c]	3.6	1.7	4.1
Shoe stores	−0.4[c]	2.1	2.5	2.9
Furniture and appliance stores	4.3[c]	2.7	2.5	3.4
Furniture and furnishing stores	4.3[c]	1.2	1.2	2.6
Appliance stores	4.4[c]	4.8	4.6	4.7
Eating and drinking places	1.1	−0.7	−0.5	−1.1
Drug and proprietary stores	6.2	0.9	1.5	−1.6
Liquor stores	N.A.	0.3	0.1	−1.0
Finance, insurance, and real estate				
Commercial banking	2.1[c]	0.6[a]	0.2	5.4[b]
Services				
Hotels, motels, and tourist courts	1.8	0.4	0.7	2.6
Laundry and cleaning services	1.7	−1.1	−1.1	0.1
Beauty and barber shops	N.A.	0.7	1.0	−2.2
Beauty shops	N.A.	1.1	0.9	−1.3

[a] 1973–1984.
[b] 1981–1984.
[c] 1967–1973.
N.A.: not available.
SOURCE: Bureau of Labor Statistics.

telephone communications, gasoline service stations, and air transportation, had very high growth rates from 1973 to 1985, ranging from 3.2 to more than 6.2 percent per year in telephone communications when the business sector as a whole was experiencing a productivity growth rate of 0.9 percent per year and the goods sector 1.0 percent per year. Other services industries, such as men's clothing stores and shoe stores, also had much higher growth rates than the business sector as a whole but were more in line with manufacturing industries, which averaged 2.4 percent per year over the period.

Some services industries, such as gas and electric utilities or laundry and cleaning services, showed very small growth rates and in some cases actual declines in productivity after 1973. In general, the range of growth rates in services industries was very broad and similar to that of goods-producing industries.

It has sometimes been mentioned that the falloff in productivity growth in the business economy of the United States is a reflection of the shift to services in the U.S. economy. Services industries are believed by some to have lower productivity growth than goods industries; thus, their greater importance over the last decade and a half is believed to have contributed to the productivity deceleration in the business economy after 1973. The wide variation in growth rates does not support that conclusion for the initial part of the slowdown period from 1973 to 1981. There was some falloff in the productivity growth rates of services industries after 1973, and this falloff contributed to the deceleration (Table 1). However, a sharp falloff in productivity growth occurred in the goods-producing sector, which until 1981 was greater than the falloff in services industries as a whole (Table 2). Since 1981, however, the picture has changed. Services industries have continued to show lower productivity growth on the average, whereas goods-producing

TABLE 2   Output per Hour, Business Sector and Goods and Nongoods Sectors, 1948–1985 (compound annual average rates)

	1948–1973			1973–1981			1981–1985		
	Q/h	Q	h	Q/h	Q	h	Q/h	Q	h
Business	3.0	3.7	0.7	0.6	2.0	1.4	1.4	3.0	1.6
Goods[a]	3.2	3.0	0.0	0.6	0.7	0.1	2.9	2.6	−0.3
Nongoods[b]	2.5	4.0	1.4	0.5	2.8	2.4	0.7	3.5	2.9
Goods − nongoods (difference)	0.7	−1.0	−1.4	0.1	−2.1	−2.3	2.2	−0.9	−3.2

[a]Goods sectors are the farm, mining, construction, and manufacturing sectors.
[b]Nongoods sectors include transportation; communications; electric, gas, and sanitary services; trade; finance, insurance, and real estate services; and government enterprises.
Q, quantity; h, hour.
SOURCE: Bureau of Labor Statistics.

industries have experienced higher growth. The services industries' slower growth has been contributing to the continued sluggishness in the U.S. productivity growth in the overall business sector since 1981.

CONCLUSION

Measurement of productivity change for services industries is not a simple task. Problems of measurement are different for each of the services industries, and the approaches followed to meet those problems vary. Each services industry has to be examined in some detail to determine whether or not reliable measures can be derived. For many services industries, adequate productivity measures can be and have been developed by the BLS.

These industry productivity indices can also be useful to companies that have introduced major technological changes. To provide some insights on the impact of various innovations on the performance of the establishments, it is desirable to develop productivity measures for these organizations. Moreover, it is helpful to have some basis for comparison. The industry productivity measures presented here can serve as benchmarks for these comparisons. However, company measures must be developed which are reasonably consistent, at least in concept, with the overall industry measure. In this way, a company productivity measure showing a rate lower than that for the industry can provide a meaningful warning signal which reflects substantive and not just measurement differences. The expansion in the number of services industries for which BLS productivity indices are available greatly facilitates this process.

Nevertheless, despite recent progress, many difficulties remain in clarifying some of the basic conceptual problems of defining the output of certain services industries, and many inadequacies are present in the data available for the measurement of productivity in this area. With regard to data, more and improved data are needed on prices, value of output, and capital input. The BLS and other government statistical agencies have undertaken initiatives to expand the coverage in the services area, and the results of these efforts should lead to data suitable for deriving more services industries productivity measures. For example, more services industries price measures are being collected, expanded hours-worked data will provide more disaggregate information, and continued BLS work on public sector productivity measurement will result in more productivity measures for this important sector.

However, some areas within the services sector, such as education, entertainment, legal and social services, and many segments of medical services, present severe conceptual and data problems for measuring productivity. Another rapidly growing area that provides great difficulties in measurement is the generation of software. Because of the complexity and heterogeneity of the products developed by this industry and the absence of data on quantities, values, or price movements, it will be virtually impossible to derive reliable productivity measures for the industry for some time. Appropriate solutions to these problems may be a long time in coming. Even additional resources may have limited effect until the conceptual problems can be dealt with appropriately. In the meantime, the BLS will continue to expand the coverage for those industries in which the problems seem tractable.

NOTES

1. This chapter expands on and updates an article (Mark, 1982) that described some of the earlier work of the BLS in developing measures for services industries.
2. In the case of multifactor productivity measures, the appropriate weight would be the sum of the factor costs and would be closer to the traditional value-added weight.
3. The literature on the problems of quality measurement for output and price indices is extensive. The approaches to developing adjustments for quality change to both types of indices are similar, but the extent to which they can be met differs. Perhaps more efforts have been devoted to obtaining adjustments for quality change to the measures for goods industries largely because of the longer history of developing output and price measures for goods industries.
4. The importance of preadmission conditions is perhaps best illustrated by a statement of Joseph A. Califano, Jr., former Secretary of the U.S. Department of Health, Education and Welfare, before the Committee on Labor and Human Resources, U.S. Senate, January 12, 1987: "Heart disease is America's number-one killer. People have the impression that coronary bypass surgery, modern cardiopulmonary techniques, miracle pills and human heart transplants are the way to battle heart diseases. Right? Couldn't be more wrong. Since 1970, our nation has experienced a dramatic 25 percent decline in deaths from coronary heart disease. The major reasons? Improved eating habits—the reduction of cholesterol—and the decline in cigarette smoking were responsible for more than half of the decline in deaths from heart disease."

REFERENCES

Becker, G. 1975. Human Capital. Chicago and London: University of Chicago Press.
Committee on National Statistics, National Research Council. 1986. Statistics About Service Industries. Washington, D.C.: National Academy Press.
Fuchs, R. 1968. The Service Economy. New York: Columbia University Press.
Gollop, F. M., and D. W. Jorgenson. 1980. U.S. productivity growth by industry, 1947–73. Pp. 17–124 in New Developments in Productivity Measurement and Analysis, J. W. Kendrick and B. N. Vaccara, eds. Chicago: University of Chicago Press.
Gorman, J. 1969. Alternative measures of the real output and productivity of commercial banks. Pp. 155–188 in Production and Productivity in the Service Industries, V. R. Fuchs, ed. New York: Columbia University Press.
Marimont, M. 1969. Measuring real output for industries providing services: OBE concepts and methods. Pp. 15–40 in Production and Productivity in the Service Industries, V. R. Fuchs, ed. New York: Columbia University Press.
Mark, J. A. 1982. Measuring productivity in service industries. Monthly Labor Review (June).
Mincer, J. 1974. Schooling, Experience and Earnings. New York: Columbia University Press.
Scott, W. R. 1979. Measuring outputs in hospitals. Pp. 255–275 in National Research Council, Measurement and Interpretation of Productivity. Committee on National Statistics. Washington, D.C.: National Academy Press.

U.S. National Center for Health Statistics. 1985. Diagnosis-related Groups Using Data from the National Hospital Discharge Survey, United States, 1982. R. Pokras and K. K. Kublishke. Advance Data from Vital and Health Statistics, No. 105. Department of Health and Human Services Pub. No. (PHS) 85-1250, January 18, 1985. Hyattsville, Md.: Public Health Service.

Waldorf, W. H., K. Kunze, L. S. Rosenblum, and M. B. Tannen. 1986. New measures of the contribution of education and experience to U.S. productivity growth. Paper presented at the annual meetings of the American Economic Association, December 28–30, New Orleans, La.

Air Cargo Transportation in the Next Economy

FREDERICK W. SMITH

The notion that the services industries of today are the cradle of the "next economy" is an accurate and important assessment: high-quality services and the advanced technologies they employ will provide the bedrock for U.S. domestic economic progress, achievement, and excellence in the twenty-first century. Similarly, U.S. services industries will play a central role in an increasingly integrated world economy.

The revolution in information technologies has captured the imagination of people everywhere. The so-called information age has inspired all kinds of predictions about the impact of instantaneous, worldwide communications. The rapid erosion of the relevance of world geography by communications technology spawned a new phrase, "the global village." During the same period, however, another industry with equally far-reaching possibilities was emerging: express air cargo transportation. Compared to the information industry, much less has been written or said about the air express industry and its impact on global economics.

In particular, few people seem to have noticed the economic significance of express carriers in important emerging management practices in distribution and logistics. The concept of just-in-time (JIT) inventory management, in which the materials of production are scheduled to arrive when they are needed in the production process, is a simple idea with enormous implications. In an economy focused on keeping inventory lean, air cargo transportation equipment and systems become "flying warehouses"—accessible, secure, 500-mile-an-hour storage facilities for those items somebody wants to use tomorrow.

The typical distribution system is paper- and people-intensive and organized to deliver goods in weeks or months. Errors, delays, and large inventories are the norm. Yet what would happen if goods could bypass traditional on-site storage, with the attendant recordkeeping and accounting, and instead be stored and shipped in a continuing process that uses computer-based express air transportation equipment? The possibilities boggle the mind. Components made in Chicago today could be on a merchant's shelf in Frankfurt, Germany, in 48 hours—routinely.

The concepts of product velocity and zero inventory are relatively new, and the ability to manage the flow of information throughout the product cycle, from purchase order to shipping documents and at all steps in between, on a global scale is just now coming of age. JIT and air express are in the forefront of that new era. If information technology created the concept of a global village, then express air transport technology is creating the concept of the global factory and the global store. The vision of a just-in-time economy is not some farfetched dream, but rather a simple step beyond where we currently stand. To understand what still needs to be done, it is important to look back briefly at what has already happened in cargo transport.

THE DEMAND FOR EXPRESS SERVICE
AND THE FEDERAL EXPRESS EXPERIENCE

Until the early 1970s, cargo fell into three basic classifications: (1) bulk and fluids, primarily agricultural or mineral products, carried by water or rail; (2) freight, including both industrial and retail goods, carried primarily by surface motor vehicles to intermediate stages in the distribution chain; and (3) letters and routine, low-priority parcels in support of, for the most part, retail trade. However, by the 1970s the need for a fourth classification began to emerge: express carriage. The express segment filled a need for transporting high-priority items for all economic sectors: commercial, government, and consumer.

Obviously, there has always been a demand for priority shipping services. In the past, pony express, clipper ships, and Railway Express have all played a role. However, with the incremental advances in electronics and microchip technology, the increasing sophistication of markets, and the economic interdependence of regions, the unserved demand for a new kind of "priority" increased rapidly after World War II and reached a peak in the early 1970s. The express segment that emerged was one in which nonperformance by the express carrier could bring severe economic consequences upon the shipper, the consignee, or both.

Federal Express developed as an innovative idea in response to these new demands for performance in the air cargo industry and the opportunities they presented. The mandate was simple: to move an item from one point to another with speed, efficiency, and predictability. Yet while the mandate was simple, its implementation was a major challenge in logistics: custodial treatment of packages, absolutely accurate records, and instantaneous tracing capabilities were new ideas in the express cargo industry.

Federal Express was quick to recognize that the new market segment would almost always demand a short-time delivery. Furthermore, the new market expectations demanded

- marked improvement in services at every point in the distribution system;
- access to, and utilization of, a highly sophisticated national and international information network;
- better logistics support systems;
- better transportation systems in general;
- state-of-the-art technologies for every critical step;
- a highly skilled and motivated work force; and
- an ability and willingness to innovate.

With the enormous growth of computerization, automation, and mechanization that began in the late 1960s, no other response would have been sufficient. The transportation landscape was changed forever: consistent and predictable overnight cargo delivery on a national—and, later, international—scale was a major breakthrough for the U.S. economic system.

Such an undertaking required a do-or-die commitment; a corporate philosophy based on people, services, and profit; and a commitment to developing and applying new technology. As a result of those commitments and its basic philosophy, Federal Express has grown from delivering a handful of packages on the first day of operation in 1973 to handling almost a million shipments daily. Its substantial aircraft fleet provides services to over 100 countries. Surface vehicles are equipped with computers that are connected to an on-line data processing system by digital radio and satellite links. In addition, Federal Express's overnight service reaches nearly 99 percent of the U.S. population.

Federal Express expects continued dramatic growth. There are several reasons for that optimism. One is that the air express industry is well-positioned to benefit from the trend toward more widespread use of just-in-time delivery. The technology and equipment are in place. It is only a matter of time before the business logistics community makes the necessary adjustments to fully implement JIT.

Another reason for the optimism has to do with the enlightened attitude toward the "services" part of the phrase "services economy." When that phrase came into vogue in the early 1980s, too many people equated a services economy with retail clerking, janitorial services, and the like. Some even feared that services would supplant manufacturing, that the United States would become a nation of people who would serve each other hamburgers or take in each other's laundry. Fast food, shoe repair, and dry cleaning are indeed services, but so are transportation, communications, health care, and finance. The rapid growth in air express and other high-value-added, technology-intensive services has given new meaning to services in our economic lexicon. In air express services, the product is convenience, efficiency, and cost-effectiveness. In this new age, with the high degree of sophistication and competition in the express segment of the transportation market, the old myths of poor pay, unskilled labor, outdated technology, and static management systems have been dispelled. Indeed, success in this market demands state-of-the-art technology in both equipment and systems, dynamic and enlightened management, and most importantly, people. Technology alone—no matter how advanced—will not bring success, but with properly motivated people who work in a stimulating environment and have a feeling of direct involvement in the firm's objectives and activities, there is no limit to what can be accomplished.

DOMESTIC AND INTERNATIONAL REGULATORY POLICY

Although the potential for both national and global gains through sophisticated express cargo operations is becoming recognized, restrictive domestic trucking and outdated international aviation treaties continue to hamper overall economic growth.

Domestic Regulation

When the U.S. Congress deregulated air cargo transport in 1977 and interstate trucking in 1980, it helped to create exceptional opportunities for risk-taking entrepreneurs in the cargo transportation industry. For one thing, competition increased.

Since airline deregulation, air express industry revenues increased almost sixfold. Today, this is a $7-billion industry and over 500 million shipments are moved in the United States each year. Between 1980 and the beginning of 1985, the number of general freight surface carriers rose from approximately 18,000 to more than 30,000, an increase of almost 70 percent. More importantly, the number of carriers authorized for nationwide general commodity services rose from zero in 1980 to 5,400 in 1987 (Rastatter, 1987).

As competition increased, prices fell and resulted in savings for both the consumer and the operator; rates fell as much as 25 percent by 1983, only six years after deregulation (Moore, 1983). Equally important, the lower prices forced the industry to become more efficient: new firms entered the market and older firms modernized their air cargo systems by buying or leasing dedicated freighters and building sorting hubs. In short, the industry became driven by customer demands rather than by regulatory dictate. Because of the lower prices and increased overall efficiency, savings to the national economy average about $50 billion per year (Delaney, 1986). So the positive effects of deregulation are not pie-in-the-sky notions—they are real.

Deregulation has also meant better services. A 1985 U.S. Department of Transportation report revealed that surface shippers received better services after deregulation than before and that the biggest improvement in the quality of services was reported by small shippers (U.S. Department of Transportation, 1985).

Intrastate Deregulation

The other side of the domestic regulatory coin is intrastate deregulation. To complete the domestic deregulation cycle what remains is to deregulate—once and for all—intrastate trucking. If the deregulation of interstate trucking spurred economic growth, it follows that intrastate deregulation would have the same effect.

A study performed for Federal Express by Gellman Research Associates estimates that substantial rate reductions can be achieved through intrastate deregulation. Some examples by state are Texas, 33 percent; Washington, 19 percent; Pennsylvania, 15 percent; Ohio, 11 percent; and New York, 10 percent. These figures suggest that there is ample opportunity to lower prices, be more efficient, and improve customer services.

No less is at stake than U.S. performance in the world market. The vast majority of goods in intrastate shipping are in reality interstate goods (and, by extension, international goods) that have been temporarily warehoused.

Air Transport Treaties and International Air Cargo Markets

The factors that created the express market here in the United States are already at work internationally. In addition, the transnational integration of production is creating pressure on existing economic, social, and industrial practices in nations trading in world markets.

It would be sheer folly to ignore these developments. Figures from the U.S. Department of Commerce show that in 1986 the value of goods moved by air transport—both exports and imports, excluding mail—increased by $10 billion and $11 billion, respectively (U.S. Department of Commerce, 1986). Moreover, a study by the U.S. Departments of Transportation and State on international air cargo markets shows that although U.S. carriers still move more air freight than airlines of any other country, they have not participated in overall market growth to the same degree as their foreign competitors (U.S. Department of Transportation and U.S. Department of State, 1988). According to that report, U.S. carriers now transport 14 percent of global air freight, down from 21 percent 10 years ago. By contrast, foreign carriers now haul 69 percent of the air commerce moving to or from the United States, up from 58 percent a decade ago.

From 1970 to 1986, the value of goods transported by air for U.S. exports increased by a factor of 10, and the value of goods imported by air transport into the U.S. marketplace increased by a factor of 20 (U.S. Department of Transportation and U.S. Department of State, 1988). The U.S. carriers are getting far less than just a smaller piece of the pie. This is true despite the fact that the United States leads the rest of the world in small-package and air express services innovations (U.S. Department of Transportation and U.S Department of State, 1988).

So while the next economy is dawning, there are significant problems still facing U.S. airlines, forwarders, airports, and aircraft manufacturers that need to be addressed if the United States is going to correct the existing competitive imbalance.

Most regulation of international air transportation is totally at odds with the demands of the express cargo marketplace. At a time when there is a critical need to recognize the different routings, physical characteristics, urgency, and schedules required for the flow of express materials, U.S. companies are hampered by archaic, bilateral air transport treaties. Such treaties were developed around the needs of passenger services with no regard for the type of new market demands inherent in air express cargo transport.

If U.S. carriers are to compete worldwide on a "level playing field," a number of significant problems will have to be resolved, but they revolve around one theme: U.S. government's responsibility to negotiate forcefully with foreign nations on trade agreements. U.S. trade negotiators need to be aware of the opportunities that the nation is missing in international trade. Once enlightened, they can then begin to focus on the real issues:

- Customs must be liberalized. Moving express documents is an integral part of a growing international market, but current country-by-country postal, telephone, and telegraph regulations are antiquated and monopolistic and so prevent the expeditious flow of time-sensitive materials. Moreover, customs operations that still focus primarily on the movement of bulk and freight—and to a lesser degree, traditional postal services—also slow the movement of goods between nations.
- Trade agreements must be reworked. Existing bilateral trade agreements are not consistent with the needs of the express cargo marketplace; treaties that focus solely on passenger services needs must be overhauled to account for the needs of the modern air cargo market in terms of route structures, physical characteristics (especially size and weight), and the urgency and schedules required for an uninterrupted flow of express materials.
- U.S. trade officials must be more actively involved in the real world of

international trade. First, the United States should constantly review and, if necessary, update its own international trade policies to keep them current with the realities of the marketplace. Next, the United States needs to implement a system of monitoring the many and varied trade policies of other nations and to negotiate changes where necessary to ensure equitable U.S. participation in the world market.

Technological change in the capability of transportation systems is creating an era in which economic opportunities abound. The express segment of the cargo transport industry is poised to make even greater economic strides once the pathway to success is cleared of outdated government-to-government agreements.

History has generally shown that nations that fail to recognize the value of trade wither and die. What is at stake here, however, is the opportunity to create jobs and the opportunity to improve prosperity both in the United States and around the world. With a global population likely to increase by more than a billion people over the next decade, the need to do so is paramount.

REFERENCES

Delaney, R. V. 1986. Digging deeper: A review of the managerial and financial challenges facing transport leaders. Transportation Quarterly (January):6.

Moore, T. G. 1983. Rail and truck reform—The record so far. Regulation (November/December):38.

Rastatter, E. H. 1987. The changing environment for motor carriers. Cited in Defense Transportation Journal (December):13.

U.S. Department of Commerce. 1986. Highlights of U.S. Import and Export Trade. Report FT990 December 70–86. Washington, D.C.

U.S. Department of Transportation. 1985. Five Years After the Motor Carrier Act of 1980: Motor Carrier Failures and Successes. Washington, D.C.: Government Printing Office.

U.S. Department of Transportation and U.S. Department of State. 1988. An Analysis of United States International Air Cargo Market, 1975–1986, Vol. I. Washington, D.C.

Changing Economics of International Trade in Services

RAUF GÖNENÇ

Services differ from manufacturing in other ways than the latter differs from agriculture. Many services employ the same technical operations, equipment, and skills as manufacturing sectors, but instead of mass-producing standard goods before marketing them, they put their productive resources into use according to the needs and specifications of their customers (tailors' services versus cloth manufacturing, computer services versus software products, mechanical engineering services versus heavy machinery production, etc.). Also, a number of services consist of the shared use of large manufactured equipment which, in alternative organizational settings, would be produced, marketed, and used as consumer durables or producers' goods (electricity-supply services versus electricity-producing equipment, air and sea cargo services versus private transportation equipment, telecommunications services versus private network supply, movie theater services versus film display equipment, etc.).

Therefore, services are not activities characterized by the use of specific techniques or the servicing of specialized markets—the two traditional ways of describing a branch of activity. They would be more appropriately defined as a special way of putting the economy's land, capital, and labor resources to use: not in the form of organizations turning out finalized goods, but in the form of organizations making them directly available, accessible, and usable by customers.

Indeed, many services are characterized by (1) the supply of a core land, capital, or labor resource: developed real estate in the case of housing services, tourist resorts, amusement parks, etc.; (2) a fixed capital stock—the telecommunications and railway networks, for example—in network services such as telecommunications and railway utilities; or (3) the specific knowledge base and skills of professionals in the professional services of physicians, lawyers, etc. Certain services are based on a more diversified combination of land, capital, and labor resources: retail banking services based on facilities, equipment, and qualified personnel at a given location; hotel services combining location in privileged sites, specific accommodation facilities, and the know-how of their personnel. What is common to all these activities and distinguishes them from traditional manufacturing is that instead of offering any final product, they supply the productive resources themselves and create value and utility through access to, and interaction with, the customers.

Seen in this perspective, the insightful chapters by Duchin and Kutscher in this volume can be interpreted as putting forward three major factors to explain the superior growth of services sectors:

1. technological progress and sophistication in capital goods, some of which have reached a level of complexity and size that makes them largely unusable (too expensive) for average private users (companies and households);

2. increasing specialization of technical skills in many areas of knowledge, making the building of pools of "state-of-the-art" human resources excessively costly or impossible for most companies and households (especially for those whose main activity is not in this given area); and

3. higher pressure for efficiency in most sectors, including government, which promotes competitive market procurement of resources as against their internally monopolistic production by users themselves.

The growth of international trade in services originates from the same factors. It is the outcome of the growth strategies of services suppliers, first in domestic and then in international markets, which build on the needs and preferences of consumers for productive resource use. It deepens the division of labor in the design and management of these increasingly sophisticated real estate, productive capital, and skilled labor inputs of modern economies.

In this chapter, therefore, the growth of international trade in services is considered a natural outcome of trends in their domestic organization. This thesis is substantiated in two sections:

• The first section presents background statistical information on the growth and structure of international trade in various services areas.

• The second section treats, in more analytical terms, the new forces and directions affecting this trade and includes some prospects for its future development.

The conclusion presents some views about the impact of the wider and freer international access to sophisticated productive resources, as allowed by international trade in services, on the functioning of the global economy.

STATISTICAL SURVEY OF INTERNATIONAL TRADE IN SERVICES

The limits of statistical information available on services trade are frequently pointed out. It is true that since they are not monitored by customs authorities, services flows are not recorded in the same systematic way as internationally traded goods. This area is certainly a long way from having the wealth of detail and precision found in the famous Organization for Economic Cooperation and Development (OECD) Trade Statistics Series—the "Red Books." Nevertheless, to the extent that services transactions generate international payments, they are captured in national balance of payment statistics. These are recompiled internationally by the International Monetary Fund (IMF) in its International Balance of Payments Annuals.

These statistics cover the so-called invisible trade transactions. They report mainly (1) the payments corresponding to the cross-the-border provision of services and (2) the factor income payments (capital and labor income) between residents of different countries. The first component covering cross-the-border trade is useful for our purposes. The second is less so, since it includes capital income originated in nonservices items such as foreign direct investment in manufacturing and financial portfolio investment. It also includes remittances of temporarily migrated workers, whatever their area of activity.

Cross-the-border trade in services is thus the only component covered systematically and relatively accurately in official statistics. Tables 1 through 4 present information on various aspects of this trade in the OECD area between 1975 and 1985. The major points emerging from this statistical review are the following:

• Cross-the-border trade in services by 10 major OECD exporters (the United States, Japan, the United Kingdom, Federal Republic of Germany, France, Italy, Canada, the Netherlands, Spain, Switzerland) increased from $120 billion in 1975 to $253 billion in 1985. This amounts to an annual growth rate of 8 percent in nominal terms over the past decade.

• The bulk of international trade in services takes place inside the OECD area. OECD countries absorbed 95 percent of total exports by other OECD countries in 1985. International services markets are located to a very large extent inside the OECD.

• The United States, France, the Federal Republic of Germany, the United Kingdom, and Japan are the major services exporters. They account for 9, 6, 6, 5, and 4 percent of total OECD exports, respectively. This group of five countries as a whole realizes 30 percent of total OECD exports.

• Sectoral breakdown of cross-the-border services trade gives the following results: 30 percent exports realized in transportation; 24 percent in tourism and travel; 9 percent in government services; 4 percent in intellectual property

TABLE 1 Cross-the-Border Services Exports by Big OECD Countries, 1985 ($ U.S. billion)

	United States	Japan	Federal Republic of Germany	France[a]	United Kingdom	Italy	Canada	Netherlands	Spain	Switzerland	Total
TOTAL SERVICES	54.1	23.4	35.5	37.5	31.1	22.7	11.5	15.2	13.2	8.6	252.8
Travel	11.7	1.1	5.9	7.9	7.0	8.8	3.7	1.5	8.1	4.1	59.8
Transportation	17.0	12.1	8.2	9.0	8.5	5.0	3.3	8.1	3.4	0.4	75.8
Government services	9.9	2.7	7.7	0.4	1.8	0.1	0.4	0.3	0.1	N.A.	22.7
Other services	15.6	7.4	13.7	20.0	13.8	8.9	4.1	5.3	1.6	4.0	94.4
Intellectual property income	5.8	0.7	0.6	0.5	1.0	1.0	N.A.	0.2	N.A.	0.1	9.9
Investment income	90.0	22.1	13.7	20.6	68.1	5.1	5.4	8.9	1.7	11.2	246.8
Labor income	0.3	N.A.	1.9	1.6	N.A.	1.7	0.1	0.4	0.1	0.4	6.5
Total services and factor income	144.1	45.5	49.2	75.1	99.2	27.8	16.9	24.1	14.9	19.8	516.6

[a]Merchant services excluded for France.

N.A. = not available.

SOURCE: International Monetary Fund (IMF), OECD.

TABLE 2 Structure of Services Exports, 1985, Billions of U.S. Dollars and Percent Distribution

	United States		Japan		Federal Republic of Germany		France		United Kingdom	
	($ billion)	(%)	($ billion)	(%)	($ billion)	(%)	($ billion)	(%)	($ billion)	(%)
TOTAL SERVICES EXPORTS	54.1	100	23.4	100	35.5	100	37.5	100	31.1	100
Tourism/travel	11.7	22	1.1	5	5.9	17	7.9	21	7.0	23
Transportation	17.0	31	12.1	52	8.2	23	9.0	24	8.5	27
Insurance	0.2	0.4	N.A.	N.A.	1.6	5	2.0	5	2.2	7
Intellectual property income	5.8	11	0.7	3	0.6	2	0.5	1	1.0	3
Construction/engineering	1.6	3	N.A.	N.A.	2.9	8	1.9	5	0.3	1
Communications	1.5	3	N.A.	N.A.	0.5	1	N.A.	N.A.	0.8	3
Consulting/technical cooperation	N.A.	N.A.	N.A.	N.A.	N.A.	N.A.	2.9	8	1.6	5

N.A. = not available.
SOURCE: IMF, OECD.

TABLE 3   Services Exports by Small OECD Countries, 1985
($ U.S. billion)^a

Austria	7.5	Norway	7.4
Belgium	13.7	Portugal	2.1
Denmark	7.6	Sweden	5.4
Finland	2.4	Turkey	2.6
Greece	2.7	Australia	4.0
Iceland	0.4		
Ireland	1.3	Total	57.1

^{a}Small countries' exports/total OECD exports = 15%.
SOURCE: IMF, OECD.

TABLE 4   Services Imports by OECD Countries, 1985 ($ U.S. billion)^a

United States	57.5 (20%)	Iceland	0.4
Japan	35.4 (12%)	Ireland	1.5
Federal Republic of		Netherlands	15.1
Germany	37.3 (13%)	Norway	7.6
France	28.5 (10%)	Portugal	1.2
United Kingdom	23.7 (8%)	Spain	5.5
Italy	17.1	Sweden	5.4
Canada	14.7	Switzerland	5.8
Austria	4.8	Turkey	1.3
Belgium	11.8	Australia	7.2
Denmark	7.5	New Zealand	1.8
Finland	2.9		
Greece	1.5		
		Total OECD	295.5

^{a}Percent of total OECD services imports are shown for the largest five importers.
SOURCE: IMF, OECD.

income; 37 percent in "other services." The last is a catch-all category, including various professional services. It has been the most rapidly growing subcategory of international services in the last decade.

• The share of different services inside the category of other services is not measurable at a global level. This detail is not available in most countries' statistics. If the sectoral distribution of U.S. exports is used as an indicator of the structure of total trade, the following results are obtained: insurance services, 0.4 percent; commissions and brokerage, 0.6 percent; films and television, 0.7 percent; construction and engineering, 3 percent; communications, 3 percent.

• Each country's services exports are characterized by specific strengths: intellectual property income in U.S. exports (11 percent of total exports); transportation in Japanese exports (52 percent of total exports, related to that country's manufacturing trade); tourism and travel in Spanish exports (60 percent of total exports); insurance in the exports of the United Kingdom (7 percent of total exports); consultancy and technical cooperation in French exports (8 percent of total exports).

• A major—and not widely known—point is that the balance of U.S. cross-the-border trade in services is negative. Deficits in subsectors such as tourism/travel, transportation, government services, insurance, communications,[1] and advertising are not balanced by smaller surpluses in intellectual property income, films and television, and construction and engineering. The net balance amounted to minus $3.5 billion in 1985.

The major shortcoming of the picture outlined above is that it ignores revenues generated by foreign subsidiaries of services companies. This is the main channel of internationalization in a number of services areas, especially for U.S. corporations. Banking and finance is the major category in which almost all foreign operations are handled by local branches and subsidiaries of U.S. institutions. Offshore banking activities in the United States are not considered domestic activities either and are not included in U.S. exports.

The Office of Technology Assessment (OTA) of the U.S. Congress has estimated revenues generated by foreign subsidiaries of U.S. services corporations to be approximately 120 percent of their direct exports (OTA, 1986). The OTA has also stated that official U.S. balance of payments statistics could seriously underestimate—in some cases by close to 50 percent—the volume of the U.S. cross-the-border trade. The bias related to foreign direct investment (FDI) is, however, not specific to the services sector. FDI is the major vector of international growth for U.S. manufacturing corporations and has created a similar inconsistency between the competitiveness of U.S. manufacturing firms in world markets and their net impact on national trade performance. The same applies equally to services.

CHANGING ECONOMICS OF SERVICES TRADE

Three Categories of Services

Three categories of services can be distinguished with regard to their potential for international trade. All services, possibly various segments of one services sector, can be distributed among these three categories:

1. The supply of capital and labor resources serves as an infrastructure to other international activities, especially trade in goods. International transportation, international communications, trade finance, and insurance are such infrastructural services. These must be purchased, practically by definition, by residents of different countries at once. The production of a service is tantamount to its trade, which means that the service is international by

its very nature and substance. International tourism is included in this category, because in all cases it involves a transaction between residents of different countries.

2. Some services are traded internationally because of the competitive advantage of suppliers. Professional services such as engineering and consultancy, computer services, accounting services, and investment advisory services, fall in this category. Exports occur because the services supplier possesses some kind of technological monopoly, a proprietary service, or a cost advantage, in the same way as for finished goods in manufacturing trade.

3. No trade can develop in "services areas closed to market competition"—be it national or international. Public utility services such as railway transportation, electricity distribution, water services, national telecommunications (in most countries), and air transportation (in most countries), as well as social services such as education, health services, and retirement insurance systems are often regulated by national governments and, therefore, closed to foreign competition. They are, in any case, not included in "marketed services sectors" as defined by OECD National Accounts.

The difference in international trade potential for each of these categories helps explain the remarkable asymmetry between the structure of domestic services output and the structure of international services flows. As Tables 1 and 2 show, international trade in services is still dominated by infrastructure-based services (i.e., international transportation, tourism, and communications), which accounts for more than 50 percent of total trade in services. In contrast this group represents no more than 5–10 percent of services output and employment at a domestic level. On the other hand, wholesale and retail trade and government services, which together represent approximately 50 percent of domestic services output in most countries, are practically absent from the international trade arena.

The major share of "international infrastructure" services in total services flows and the negligible role of government-related services are natural and do not need much explanation. What may be striking, however, is the very limited exposure to international competition of most of the marketed business and household services.

Three major factors account for this limited international competition in marketed services:

1. In a number of services areas, *needs and demand* are structured by local business and household culture, and can be satisfied by hardly anybody except local, long-established services suppliers. Japanese domestic trade systems, German banking, U.S. corporate finance services, and French restaurant sectors are examples of market areas in which not only cross-the-border trade, but also foreign services supply through local investment and establishment, are very difficult due to the specifics of local demand. Demand

for productive resources (which are often customized) is more qualitatively differentiated at the international level than demand for finished, standardized, manufactured goods.

2. International supply of services is often more *complicated and costly to design* than international manufacturing production. Business organizations in services are "open" systems interacting with customers, and their internationalization involves more sophisticated organizational structures and human resources bases than most manufacturing activities. As a result, the basic investment and organizational design necessary to sell services internationally are often larger, and imply deeper changes for a company, than those necessary for exporting goods or transferring (delocalizing) standardized manufacturing techniques.

3. There is the question of *incentives*. In many sectors, services firms are of very small size, frequently family based, and in most cases private organizations. Their business strategies are rarely rate-of-return maximization or growth oriented; they are rather shaped to maximize private income for owners. This is frequently observed in low-technology, single-shop services activities, as well as in most sophisticated professional services, such as those of physicians, lawyers, and engineering firms. As long as these operations remain personal and private organizations, they have little incentive to consider change for better valuation and market capitalization of their assets. They rarely have objectives and prospects for widening their output and market share, nationally or internationally. Only exceptional individuals give impetus for organizational growth and change in such "small organization" sectors (Heskett, 1986).

These three characteristics of marketed services are currently undergoing important changes that are exposing many sectors to new forces of competition, growth, and internationalization. In the following sections the most significant aspects of this new trend are discussed.

More Competition in the Supply Side

Many services industries are undergoing deep restructuring in their competitive environment: the bulk of new entrepreneurial talent in OECD countries, as well as capital resources in search of good investment opportunities, tends today to seek out services areas as an alternative to saturated manufacturing activities. Notably, interest from financial investors and entrepreneurs includes prospects for takeovers, mergers, and acquisitions among private services companies, which leads to new pressure and opportunities for a more active use and market capitalization of the business assets of these companies.

At the same time, in traditional manufacturing sectors, competitive rivalry encourages firms to develop services activities such as marketing and retail trade, consumer credit and insurance, consulting, and engineering. This sometimes incites such sectors to compete directly with established services providers (e.g., computer hardware manufacturers diversifying into computer services and software, or car manufacturers diversifying into car credit and insurance).

New technological opportunities foster the new competitive environment. In particular, possibilities offered by information technology in retail and wholesale trade, financial and insurance services, engineering, design and consultancy services, etc., offer new channels for productive resource supply, services differentiation, and competitive advantage building, nationally and internationally.

The Case of Financial Services Financial and banking services are particularly illustrative of the depth of these changes in the supply-side and competitive policies in services industries. Market environment has become much more difficult, following government deregulation policies in practically all OECD countries, which liberalized interest rates and cross-entries between market segments. These policies have also promoted, to a lesser extent, foreign establishment and supply of financial services. The respective roles of government policies and of purely private pressures in increased competition are difficult to isolate from each other; the whole process has simply led to a much more competitive environment.

In retail banking, the new environment has led to price competition (increase in deposit interest rates) and to the marketing of new services (interest-bearing call accounts, credit and cash card services). In stock brokerage, intermediation margins came down in accordance with costs, reflecting notably the economies of scale in large-volume transactions, and services diversified greatly with increased choice in terms of market information and analysis available. In the corporate finance area, price competition reduced underwriting fees, and new services have been introduced to meet the demand for corporate cash management, interest and currency-risk hedging, and merger and acquisition financing.

Banking and financial services are becoming less standardized and more sophisticated throughout this process. Most financial institutions have invested huge amounts in information technology applications to reduce costs and diversify services, as in automated teller machine (ATM) networks, back- and front-office automation, electronic fund transfer systems, trading rooms, and global information networks. The capital (equipment) and labor resources put at the disposal of financial services customers are, therefore, undergoing considerable widening and sophistication. Financial institutions are also enriching their services through internal personnel training and external hiring of more highly skilled specialists.

Competitive differentiation of resources (services assets) by financial companies has created scope for their wider capitalization at the international level. Most of the services technologies (including organizational know-how and centrally integrated information networks) are hardly transferable among organizations. Internationalization then has to take the form of the multinationalization of financial companies, instead of external growth mechanisms such as technology transfer agreements or licensing.

Measurement of the growth of international activities by financial services firms creates a wealth of technical problems (see OTA, 1986). The OECD is currently in the process of designing a new statistical system for this purpose (OECD, 1987a). In all cases, the growth of international activities and competition by these firms has developed tremendously during the 1980s. In one instance, international assets and liabilities of OECD commercial banks evolved as described in Table 5.

In cases where financial innovations are easy to reproduce—however radical they may be—innovator organizations try, though not always very convincingly, to protect their innovation on intellectual property grounds (Merrill Lynch trying to patent its cash management account system, for example). However, the national and international market finds ways to respond quickly in imitating and diffusing successful new services (Levich, 1987). Competitive innovation is becoming increasingly difficult under these conditions, requiring the building of more and more sophisticated and, as far as possible, "company-specific" physical assets, organizational design, and labor resources.

Perceived as a quasi bureaucracy ("gray flannel suits") or a gentlemen's club ("borrow at 3 percent, add 3 percent margin, play golf at 3 p.m.") only a decade ago, financial services are today being transformed into the archetype of a relentlessly innovating and fiercely competitive Schumpeterian industry. The growth of their international markets and the rise of global competition in this sector are the outcomes of this transformation.

More Sensitive and Responsive Demand Side

Services markets are undergoing changes on the demand side as well. Facing a much more aggressive competitive environment, firms from all sectors search to minimize the costs of the services they secure and tend to look at external procurement possibilities as an alternative to the services they produce internally (Ergas, 1983). This is seen in low-technology activities such as office cleaning, cafeteria services, and printing-copying services, as well as in high-technology activities such as computer services, strategic auditing, and research and development.

Governments, including local governments, and public agencies also participate in this movement of rational services procurement. Facing increasingly tighter budgetary constraints, they begin to "farm out" a growing share

TABLE 5   Internationalization of Banking Assets and Liabilities (foreign assets and liabilities as percentage of total assets and liabilities)

	Assets (%)		Liabilities (%)	
Country	1970	1981	1970	1981
Australia[a]	—	0.6	—	1.1
Austria[b]	10.7	24.5	9.8	27.8
Belgium[c]	33.4	57.8	39.0	68.7
Canada[a]	19.8	17.3	14.3	27.1
Denmark[c]	6.7	29.1	7.0	28.1
Finland[b]	4.4	11.5	5.6	17.5
France[b]	15.9	33.7	17.0	32.3
Federal Republic of Germany[b]	8.8	10.2	5.6	8.1
Greece[c]	3.5	7.8	4.7	22.0
Iceland[b]	1.0	2.9	2.7	17.2
Ireland[c]	36.0	47.1	29.8	49.2
Italy[b]	12.6	12.6	12.6	15.9
Japan[b]	3.7	6.6	3.1	7.9
Luxembourg[b]	84.2	97.5	57.5	90.4
Netherlands[b]	27.0	39.8	25.9	39.2
New Zealand[a]	7.1	7.0	1.1	2.3
Norway[b]	7.4	6.0	5.5	10.9
Portugal[b]	5.6	7.7	0.8	27.7
Spain[b]	3.5	8.5	4.2	14.9
Sweden[c]	7.0	9.7	5.4	18.2
Switzerland[d]	33.7	50.1	28.9	42.8
Turkey	—	4.7	—	0.3
United Kingdom[b]	46.1	67.9	50.2	69.9
United States[c]	2.6	15.1	6.2	11.3
Total OECD	12.1	23.7	11.3	23.4

NOTE: Data are not fully comparable across countries.
[a]Chartered or trading banks.
[b]All deposit money banks.
[c]Commercial banks.
[d]All banks including trust accounts. Balance sheet data include domestic interbank deposits.
SOURCE: OECD, Internationalization of Banking Study (1983).

of their services activities. Given the size of government expenditures in OECD countries (from 25 to 45 percent of GNP), this movement has, and will continue to have, a major competitive impact on a number of services markets. For example, the role of government procurement in computer services has been an important, yet still not widely known, factor in the growth of the computer services and software industry (OECD, 1985).

Internationalization of services markets comes as a direct consequence of this environment. Aggressive firms willing to widen their market share face clients ready to revise their long-term and stable procurement relationships with their traditional services providers, including their own internal services departments.

The Case of Financial Services   The procurement of financial services by companies, households, and governments is again a relevant case here. Facing strong and uninterrupted competitive pressures since the 1970s and taking advantage of the structural changes in financial markets, numerous enterprises have renewed their financing patterns in order to reduce capital costs nationally and internationally. Direct borrowing from capital markets through security issues and an increase in the number of, and competition between, banking partners of each company have been major trends. Borrowing on Euromarkets by big industrial corporations, where regulatory "shadow costs" as well as underwriting fees are minimal, has been the international outcome of this trend.

In retail financial services, negative real interest rates offered by deposit-taking banks in the inflationary environment of the 1970s have led many households to diversify their financial portfolio and savings patterns. A wider use of nonbank financial intermediaries, including new money-market and mutual funds, as well as the use of recently introduced and more rewarding banking services (interest-paying call accounts, individual retirement accounts, etc.) have been the major trends. Development of consumer telebanking (home banking) services is still marginal in most countries, but the emergence of similar telebrokerage services may increase the direct role of individual investors in security markets. Supported by technical analysis and decision-support software offered by telebrokerage companies, such growth would represent a significant shift in private investing, which has been characterized by progressive institutionalization over the past decades.

Finally, government financing is also characterized by increasing sophistication of borrowing techniques and public debt instruments used, as well as by fiercer competition between financial intermediaries serving government treasury needs. The use of auctionlike (competition) techniques in the fixing of underwriting margins in public issue flotation and the increased recourse to international borrowing by many governments, which often implies procurement of services from international financial institutions, are the major trends. In certain countries, governments and big public corporations even play the leading role in the generalization of innovative and competitive borrowing tactics.

Trade in other professional services, such as corporate insurance services, accounting and auditing services, and computer services, also develops on this basis. Domestic markets in these services have developed considerably since the 1970s as a result of more active and more assertive corporate redeployment and reorganization policies (among users). Consequently, this group has also been the most rapidly growing services category in international trade. The volume of U.S. exports of professional services increased by more than 100 percent from 1975 to 1985. Their share in total services trade remains limited; they reached approximately 10 percent of U.S. exports in 1985.

New Forms of Internationalization

Internationalization of formerly local services is taking various organizational forms. In addition to classical cross-the-border trade and sales by local branches or subsidiaries, certain original forms of internationalization are being developed. These new forms represent new and more efficient ways of diffusing services know-how internationally, and some of them are also seen in manufacturing activities.

Telecommunications make possible the delivery of certain services (which previously required face-to-face interaction between customers and the services supplier) through telecommunications lines. This applies especially to cases in which the substance of the services consists of information and messages (numerical, textual, voice, or images). Television broadcasting was an ancestor of this kind of service. Data bases, telesoftware, telebanking for corporations and households, on-line airline and hotel or motel reservation systems—all are services that have developed recently through telecommunications lines (Bruce et al., 1987). They will grow and diversify further in the future, probably into such areas as interactive educational programs and remote health diagnosis.

The decrease in telecommunications costs is very relevant, as distance becomes a lesser factor. If international telecommunications tariffs are adapted to the emerging cost structure in telecommunications, teleservices should become more easily available on international markets (OECD, 1986b). On-line data bases have been the pioneer of such international teleservices.

Intermediate cases are observed when interaction between parties can be standardized progressively and part of the transactions implemented at a distance. This is seen, for instance, in mail-order retail trade and on-line accounting services. Part of the information necessary for these transactions is standardized and can be exchanged at a distance through telecommunications lines.

In other cases where face-to-face interaction with local consumers requires experienced local personnel (i.e., where the internationalizing firm cannot develop the necessary skills internally), certain hybrid forms of investment occur. Joint ventures with local firms, mergers and acquisitions, partnerships, etc., are becoming the dominant form of internationalization in such skill-intensive services. Internationalization in investment banking, investment advisory services, venture capital investing, accounting and auditing services, management consultancy, and advertising have frequently taken the form of partnerships over the past decade.

Also, in several skill-intensive activities, the optimal organization for a competitive firm is one that gives direct incentives to its key personnel (Peters and Waterman, 1985). In services industries, a related organizational innovation has been "franchising." This is a technique of licensing proprietary services technology, in which the core firm (franchiser) supplies key inputs to the franchisee through long-term contracts and allows use of its brand name under restrictive conditions. Extensively applied in several services industries over the past decade, franchising diffuses quickly at the international level. It has become the dominant form of international growth in sophisticated consumer services areas such as hotel services, fast-food distribution, and fashion retail shops.

The Special Case of Public Utility and Social Services

A final category of services may present major internationalization opportunities for specialized firms: areas that were organized as public monopolies or regulated industries until the 1980s, but are being transformed into competitive "marketed services" sectors as a result of new governmental policies (OECD, 1988).

Deregulation, competition, and in many cases privatization constitute a major thrust for new structural policies in OECD countries. Such policies have been designed with a wealth of expectations, and with the goal of making monopolistic and regulated sectors more efficient and more responsive to economic needs. These sectors generate a substantial share of OECD gross national product (GNP), and new policies have already been applied to air and road transportation, telecommunications, and gas and water distribution.

More controversial organizational reforms are on the way in the area of social services, such as health care, pension insurance, and education and training. Significant experiences concerning market supply of some of these services should develop in a number of countries over the next decade (OECD, 1988).

This movement will open new internationalization opportunities in the services industries concerned. First affected will be infrastructural services that are already supplied internationally (category 1, above), yet remain under intermonopoly and cartel arrangements. This movement has already begun in telecommunications trade, air transportation, and sea cargo transportation (Peat Marwick Mitchell and Co., 1986).

Penetration of international competition into domestic deregulated sectors (domestic telecommunications, transport, energy production, etc.) will be slower. Most governments consider these activities as being of national and social interest, and may be reluctant to "lose control" over them.

This had most often been the reason behind the initial nationalization or regulation of these industries. However, even in these cases, internationalization can develop more quickly than foreseen if it takes smoother forms such as joint ventures, technology transfer agreements, franchising, or partnerships.

Rapid technological change in newly deregulated sectors may strengthen the need and scope for internationalization. New air transportation techniques (hub systems), telecommunications services through satellites and fiber optics, value-added telecommunications services, highly capital-intensive health care technologies, etc., not only will widen consumer demand for novel services, but also will improve suppliers' ability to compete internationally.

Similar developments can be expected in the long run in social services. It is clearly difficult, particularly for a European, to imagine such services becoming subject to international competition. Nevertheless, this is most probably the direction of technological and organizational development in education, health care, and retirement insurance. Expenditures for these services represent a large market (more than 20 percent of GNP in most countries), and their internationalization would be a major structural challenge for their established suppliers.

CONCLUSION

What, basically, does the development of trade in services mean? It means that an increasing amount of land, capital, and labor resources in modern economies supplies services through the market, instead of being bought (or hired in the case of labor resources) by single customers and put into action as internal inputs exclusively used by one specific organization.

A nonnegligible share of real estate, capital, and labor resources is already accessible through services markets in manufacturing or indeed even in agricultural economies. This has been, for example, the case in hostel services, ship and railway transportation, and physicians' services. The cost of building these resources, as opposed to their low frequency of use by single customers, has already structured the economics of their supply in the form of marketed services for a long time.

However, it is with the recent technological sophistication, differentiation, and increase in the size and cost of productive resources that directly marketed services are becoming a wider form of economic organization. In areas where it applies, this represents a reversal of the trend toward the internalization of capital, labor, and real estate resources by companies and households, via purchases of consumer durables and proprietary industrial equipment, hiring of specialists as members of staff, and building or acquisition of private housing and private office buildings.

Obviously, not all productive resources are becoming available on the market, and companies are not being transformed in asset terms into hollow organizations that only coordinate the supply of complementary services by external asset owners. They are, however, concentrating their assets in their core business areas and attempting to build unique, competitive, and company-specific asset bases there, while delegating to the market the supply of more peripheral services for their activities. Along the same lines, in the household sector, not all household assets are being liquidated to be replaced by occasional services procurement, but there are several indicators of a higher reliance on the services of externally owned assets beginning with the housing area.

The growth of international trade in services in this context means that business organizations and households in one country are now gaining a larger access to productive resources available in other countries. The fact that a Turkish electronics company can now solicit the engineering services of a California microchip design company, that a Korean shipbuilder can build on the leasing and other financial/marketing services of a British merchant banker, or that a Bolivian leather manufacturer can possibly use the international marketing services of a Japanese firm in all instances means that resources previously limited to the use of a single company, region, or nation tend to become available for much wider international access and use when they are supplied as services on markets.

The implications of this trend for lowering a number of barriers to entry and mobility in many industries are obvious. This allows potential entrants and various competitors to concentrate their efforts on the essential assets of an industry, and not be handicapped by the possible underdevelopment of their resources, or the resources of their national economy, in related but strategic peripheral areas. The case is especially strong when extremely sophisticated, more and more indispensable, but at the same time noncentral support services are concerned, such as insurance, finance, marketing, engineering, and value-added telecommunications services. In all these areas, the development of international trade in services lowers part of the barriers to entry in industrial competition for firms from all countries.

The growth of international trade in services is certainly not achieving the ultimate field-leveling role in global competition. It does not affect the core strategic resources of various industries, while it is indeed around the mastery and the differentiation of those very resources (assets) that global competition is being shaped. Moreover, many services markets and firms are still rather regional in character and scope, are not even nationally integrated, and consequently do not yet really guarantee access by international customers

(OECD, 1987c). Regulatory barriers to trade, where they apply, also constitute obstacles in this direction.

Nevertheless, increased international trade in services represents a major trend for more global and effective competition in the world economy. This impact should become clearer in the future, especially as more Third World enterprises learn to take advantage of, and build on, the availability of productive and support resources to which they now have access.

ACKNOWLEDGMENTS

This chapter presents the personal views of the author and does not commit the OECD. Special thanks are due to Marie-Pierre Faudemay, from the OECD's Trade Directorate, who gave valuable advice. Thanks are also due to colleagues Derek Blades, Sarah Chapman, Henry Ergas, Bruce Guile, Mufit Cinali, Dani Rodrik, and Barrie Stevens who commented on an earlier draft.

NOTE

1. This surprising deficit in communications should be related to the current payment system in international telecommunications. Imbalance between incoming and outgoing calls to and from the United States implies compensating payments to foreign telecommunications monopolies.

BIBLIOGRAPHY AND REFERENCES

Aronson, J. D., and P. F. Cowhey. 1984. Trade in Services: A Case for Open Markets. Competing in a Changing World Economy Project. Washington, D.C.: American Enterprise Institute for Public Policy Studies.

Bank for International Settlements. 1986. Study Group of Ten Countries. Recent Innovations in International Banking. Basel.

Bhagwati, J. 1987. Trade in Services and the Multilateral Trade Negotiations. The World Bank Economic Review 4:549–569.

Blades, D. 1987. Goods and services OECD countries. OECD Economic Studies (Autumn):160–183.

Brender, A., and J. Oliveira-Martins. 1984. Les changes mondiaux d'invisibles: Une mise en perspective statistique. Paris: Economie Prospective Internationale.

Bressand, A. 1986. International division of labour in the emerging global information economy: The need for a new paradigm. Promethee (October).

Bruce, R. J., J. P. Cunard, and M. D. Director. 1987. Telecommunications and transactional services: A case study of emerging structural and regulatory issues in the financial services sector. London: International Institute of Communications.

Commission of the European Communities. 1986. Europe and the future of financial services—Proceedings of a symposium. Brussels.

Congressional Budget Office, U.S. Congress. 1987. The GATT Negotiations and U.S. Trade Policy. Washington, D.C.: U.S. Government Printing Office.

Duchin, F. 1988. Role of services in the U.S. economy. In Technology in Services: Policies for Growth, Trade, and Employment. Washington, D.C.: National Academy Press.

Ergas, H. 1983. Corporate strategies in transition. In Industrial Policies in Europe, A. Jacquemin, ed. New York: Oxford University Press.

Gönenç, R. 1986. Changing Investment Structure and Capital Markets. Brussels: Center for European Policy Studies.

Heskett, J. L. 1986. Managing in the Services Economy. Cambridge, Mass.: Harvard Business School Press.

Inman R. P., ed. 1985. Managing the Service Economy—Prospects and Problems. Cambridge, England: Cambridge University Press.

Kakabadse, M. A. 1987. International Trade in Services—Prospects for Liberalization in the 1990s. Paris: Atlantic Institute for International Affairs.

Kravis, I. B. 1985. Services in world transactions. Pp. 135–160 in Managing the Service Economy—Prospects and Problems, R.P. Inman, ed. Cambridge, England: Cambridge University Press.

Krommenacker, R. J. 1986. The impact of information technology on trade interdependence. Journal of World Trade Law (July–August):381–400.

Kutscher, R. E. 1988. Growth of services employment in the United States. In Technology in Services: Policies for Growth, Trade, and Employment. Washington, D.C.: National Academy Press.

Levich, R. M. 1987. Financial innovations in international financial markets. National Bureau of Economic Research, Working Paper No. 2277 (June).

McCulloch, R. 1987. International competition in services. National Bureau of Economic Research, Working Paper No. 2235 (May).

Office of Technology Assessment, U.S. Congress. 1986. Trade in Services: Exports and Foreign Resources—Special Report. Washington, D.C.: U.S. Government Printing Office.

Organization for Economic Cooperation and Development. 1983. Internationalization of Banking: The Policy Issues. Paris.

Organization for Economic Cooperation and Development. 1984. International Trade in Services: Banking. Paris.

Organization for Economic Cooperation and Development. 1985. Software: An Emerging Industry. Paris.

Organization for Economic Cooperation and Development, Center for Educational Research and Innovation. 1986a. The Evolution of New Technology, Work and Skills in the Service Sector. Paris.

Organization for Economic Cooperation and Development. 1986b. Changing market structures in telecommunications services. Proceedings of an OECD Symposium. Amsterdam: North Holland Press.

Organization for Economic Cooperation and Development. 1986c. Venture Capital in Information Technology. Paris.

Organization for Economic Cooperation and Development. 1987a. Elements of a Conceptual Framework for Trade in Services. Paris.

Organization for Economic Cooperation and Development. 1987b. International Trade in Services: Securities. Paris.

Organization for Economic Cooperation and Development. 1987c. Internationalization of Software and Computer Services. [restricted]. Paris.

Organization for Economic Cooperation and Development. 1987d. Member Countries' Data on Trade in Services. [restricted]. Paris.

Organization for Economic Cooperation and Development. 1987e. Ownership Linkages in Financial Services. [restricted]. Paris.

Organization for Economic Cooperation and Development. 1987f. Tables on International Trade in Financial Services. [restricted]. Paris.

Organization for Economic Cooperation and Development, Center for Educational Research and Innovation. 1987g. Changing Technology, Skills and Skill Formation in the Service Sector—A Report on the Experience in Financial Service Firms in Five OECD Countries. [restricted]. Paris.

Organization for Economic Cooperation and Development. 1988. Structural Adjustment and Economic Performance. Paris.

Peat Marwick Mitchell and Co. 1986. A Typology of Barriers to Trade in Services. Brussels: Commission of the European Communities.

Peters, T. J., and R. H. Waterman. 1982. In Search of Excellence: Lessons from America's Best Run Companies. New York: Harper & Row.

Quinn, J. B. 1987. The impacts of technology in the services sector. Pp. 119–159 in Technology and Global Industry, B. R. Guile and H. Brooks, eds. Washington, D.C.: National Academy Press.

Sapir, A. 1986. Trade in investment-related technological services. World Development 14(5).

Saxonhouse, G. 1985. Services in the Japanese economy. Pp. 53–83 in Managing the Service Economy—Prospects and Problems, R. P. Inman, ed. Cambridge, England: Cambridge University Press.

Shelp, R. 1981. Beyond Industrialization: Ascendancy of the Global Service Economy. New York: Praeger.

Stalson, H. 1985. U.S. Trade policy and international service transactions. Pp. 161–178 in Managing the Service Economy—Prospects and Problems, R. P. Inman, ed. Cambridge, England: Cambridge University Press.

Thomas, D. R. 1978. Strategy is different in service businesses. Harvard Business Review, 56(4):158–165.

International Trade in Financial Services: The Japanese Challenge

RICHARD W. WRIGHT AND GUNTER A. PAULI

Little more than a decade ago, the Japanese took the world by surprise with a frontal assault on world consumer electronics, automobile, motorcycle, and photographic markets. Those and other manufacturing industries in North America and Europe are still reeling from the Japanese onslaught, as the world seeks to better understand the scientific and managerial technologies with which Japanese firms catapulted so quickly to dominance.

Today the Japanese are again on the move, and a second wave of Japanese competition is about to hit the western world. This time the target is services, and financial services in particular. Once again it appears that the West will be taken largely by surprise.

How have the Japanese moved so quickly to the forefront of services sectors long considered the exclusive domain of westerners? Does analysis of the strategies they used so successfully in manufacturing sectors help us to understand their current competitive thrusts in services? What organizational technologies do they have at their disposal? What sorts of responses are appropriate to the new international competition in services? This chapter addresses these important questions.

THE DIMENSIONS OF SUCCESS

Japan's sudden emergence at the forefront of international financial services is evident in many ways. As examples:

- The Japanese are now richer than Americans. Japan's average per capita income topped $17,000 near the end of 1986, compared with $16,000 in the

United States (*The Economist*, 1986a). They also save much more than others (*The Economist*, 1986b): 17 percent of their disposable income, on average, versus some 4 percent in the United States (Figure 1).

• Japan is the largest exporter of capital the world has ever seen. Due to the combination of high domestic savings and large foreign trade surpluses, Japan has exported almost $400 billion worth of long-term capital over the last three years, far more even than the Organization of Petroleum Exporting Countries (OPEC) in its heyday (Figure 2). And unlike the OPEC countries, which recycled their excess funds through deposits at major foreign-owned banks throughout the world, the Japanese resolutely funnel their international financial flows almost exclusively through Japanese financial institutions, thus providing Japan's banks and securities companies with a massive reserve of resources for their global strategy in world financial markets.

• Tokyo has become the world's leading financial center. The Tokyo Stock exchange has more volume already than all the exchanges of Europe combined; it recently topped even New York as the world's largest exchange, with a market capitalization of $2.9 trillion versus New York's $2.7 trillion (*Echo de la Bourse*, 1987). Even the comparatively minor Osaka exchange has soared past London to become the world's third largest, after Tokyo and New York (*Time*, 1986).

• The seven largest commercial banks in the world are all Japanese and 13 Japanese banks rank among the largest 25 (Figure 3). This has all come about extraordinarily quickly. As recently as 1980 none of Japan's banks ranked among the world's top five in size of deposits, and only one (Dai-Ichi Kangyo) ranked among the first ten.

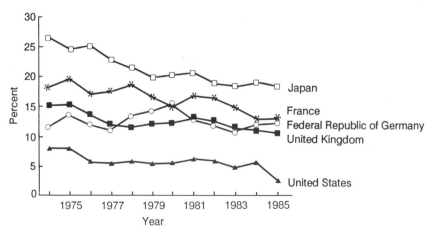

FIGURE 1   Personal savings as percent of disposable income.
SOURCE: *The Economist* (1986).

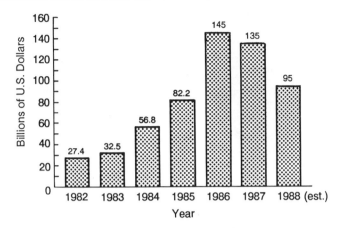

FIGURE 2   Japanese long-term capital outflow, 1982–1988. SOURCE: Bank of Japan.

However, growth in size is only one part of this remarkable success story. Japanese banks are capturing overseas markets in much the same way their manufacturing counterparts did earlier. They recently nudged out Americans as the world's largest international lenders, with $650 billion in loans outstanding, or nearly one-third of the total worldwide assets of banks (Figure 4). Their dominance in international lending shows no sign of abating. A recent report predicts that by 1990 Japan will have an astonishing $1 trillion in loans outstanding (*Euromoney*, 1985). In addition to seizing major chunks of American and European commercial lending, Japanese banks are expanding very rapidly into new activities abroad such as leasing, trust banking, and securities underwriting.

• The four largest securities companies in the world are Japanese. Like their banking counterparts, Japanese securities companies have become increasingly dominant forces in their international activities. The "Big Four" Japanese houses (Nomura, Daiwa, Nikko, and Yamaichi) now rank as the world's largest. Nomura Securities Company, largest in terms of assets, is bigger even than the American giant Merrill Lynch, and vastly more profitable: its pretax profits for 1987 exceeded $3 billion. So huge is Nomura, in fact, that its market capitalization, or what the stock market says all its shares are worth, is ten times that of Merrill Lynch (Hector, 1986).

Japanese securities firms are aggressive at home and abroad, both in expanding their securities business and in rapidly innovating into new product areas traditionally the turf of other financial institutions. They are becoming increasingly adept at using their financial clout to cut prices in U.S. and European underwritings, taking large chunks of new issues back to Japan,

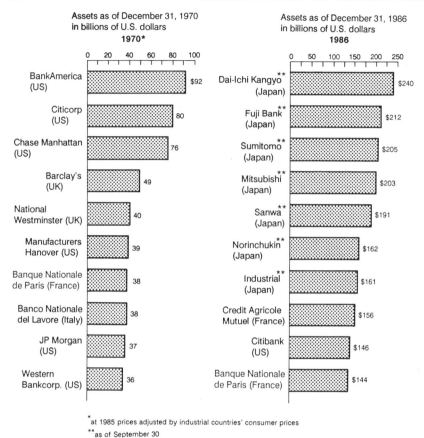

FIGURE 3 The world's largest commercial banks. Banks ranked by deposits in billions of U.S. dollars as of December 31, 1986. SOURCE: *American Banker*, company reports, *Fortune*.

where their placement power enables them to dispose of their holdings quickly (for a recent account of their placement ability and diversification strategies, see *Time*, 1988). Their profitability at home also allows them to expand into other lucrative activities abroad, as they embark on a path that will eventually make them truly global financial supermarkets.

• Japanese services firms are poised to become major players in other industries as well. Although attention in the West is focused at the moment on the dramatic rise of Japan's banks and securities companies, the Japanese are becoming major world competitors in other services areas also. For example, the world's largest transportation company (NYK) is Japanese, as

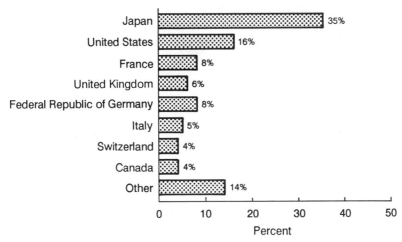

FIGURE 4 Worldwide assets of banks by ownership of banks (September 1987). SOURCE: *The Economist*; Bank for International Settlements, January 1987.

is the largest life insurance company (Nippon Life), the largest tourism organization (Japan Travel Bureau), and the second-largest advertising agency (Dentsu). The successes of the Japanese banks and securities companies may be only the leading edge of impending Japanese competition in a much broader range of services.

ELEMENTS OF COMPETITIVE STRATEGY

Head-on global competition in the services industries is a relatively new phenomenon, with little in the way of reliable models or precedents. Traditional models of industrial structure and manufacturing competition have limited relevance to service sectors, where the types of inputs and the nature of productive activity differ so from those of manufacturing.

The Japanese financial services sector, however, has embarked on a strategy remarkably parallel to that which earlier placed Japan in the forefront of information technology and other key markets. Tracing their strategy for entering and penetrating the information technology market can provide useful insights into their current competitive challenge in financial services.

Their overall approach may be characterized succinctly as the Termite Strategy. The Termite Strategy invokes no master plan, and no single leader to orchestrate the steps. Yet, just as termites act independently and create a significant common result, the Japanese financial services houses are all driving toward common objectives, each one seeking to maximize its strengths and minimize its weaknesses to the greatest extent possible, searching out the areas of greatest demand and paths of least resistance from competitors.

Under the Termite Strategy there is no big thrust for the target, no single front line, but thousands of small steps.

The purpose of this section is to analyze those steps that the Japanese are and will be taking to dominate the market for financial services, through a comparison with their successful strategy in attacking the information technology market. The paths are remarkably similar, and the consequences for unwary U.S. and European competitors may be equally unfortunate.

Elements of the typical Japanese strategy may be summarized generally as follows:

1. Japanese firms identify segments of the targeted industry and choose an appropriate entry segment. They focus intensively on that targeted sector, flooding the market with goods or services of high quality and low price.

2. They join with foreign competition in other segments of the industry, where foreign know-how or technology is superior to that of the Japanese. These joint ventures enable Japanese firms to gain important expertise in their own market before competing head-on in foreign markets.

3. They build up a strong distribution infrastructure.

4. After having established themselves securely in foreign markets through high quality and low prices, Japanese companies are accused of "dumping." Protectionist political pressure from foreign governments allows them to raise prices. Higher prices mean higher profits to Japanese firms, as well as to local competitors, at the expense of the consumer.

5. Surplus profits from increased prices provide the capital for Japanese firms to manufacture or operate abroad rather than exporting Japanese products or services. The Japanese ensure that high value-added operations remain at home in Japan, with mainly assembly-type operations abroad. Profits and expertise flow back to Japan, while low value-added jobs formerly lost to Japanese competition are reinstated abroad under the umbrella of Japanese investments.

6. Finally, the Japanese engage in takeovers to rescue failing companies abroad, and finance research and scholarships at leading western universities to secure access to well-trained young graduates.

Table 1 summarizes the strategy and compares its successful application by the Japanese in information technology to the approach being taken in financial services today. This table captures succinctly the main elements of strategy discussed below.

THE JAPANESE STRATEGY IN INFORMATION TECHNOLOGY

Reconnaissance

An appropriate target entry segment for any new market, from the Japanese perspective, would be one in which the financial resources of the competition

TABLE 1   Japanese Business Strategies: The Comparison of Information Technology and Financial Services

Information technology

1. Japanese firms enter the information technology market through high-volume price-sensitive semiconductor segment where competition has least resistance due to their focus on financial returns instead of market share.

2. Japanese firms move into supplier contracts and original equipment manufacturers (OEM) agreements penetrating other segments of the information technology market such as personal computers, peripherals, office equipment . . . to acquire manufacturing expertise.

3. Firms first secure their market through independent distributors, then establish direct sales channels centrally controlled by the Japanese headquarters.

4. After acquiring a major market share Japanese firms are accused of "dumping." Political agreements safeguard their market position and raise their profitability while western competitors feel that the "steamroller" has been stopped. The Japanese market will be "opened" for sophisticated chips the Japanese do not yet manufacture.

5. Japanese establish joint ventures and invest in new assembly lines, making sure that high-value-added operations remain in Japan, while overseas profits increase in relative importance. They also build the largest research and development institutes in the world.

6. Japanese firms consolidate their leadership position by rescuing bankrupt manufacturers, often at the demand of western governments, and even engage in hostile takeovers. They heavily finance fundamental research in artificial intelligence at leading western universities and channel the results back to Japan.

Financial services

1. Japanese firms enter the financial services market through the government lending segment where margins are crucial and the governments' appetite seems insatiable. Huge balance of trade and savings are channeled to foreign governments, creating political goodwill.

2. Japanese firms join with financial houses in other segments of the financial services market, such as foreign exchange, trust and retail banking, leasing, to acquire financial engineering expertise.

3. Firms build up large infrastructure through representative offices and branches. Offices are managed by Japanese staff under direct supervision of Japanese headquarters.

4. After acquiring a leading market share, Japanese firms will be accused of "dumping." Political deals will "open" the Japanese financial markets, but these will be either saturated (e.g., banking) or where foreign expertise is most needed.

5. Japanese financial houses obtain banking licenses, primary dealer status, seats on the New York Stock Exchange and London Stock Exchange, and dominate the minor financial centers, all through bilateral deals. Profit and expertise flow back to Japan. They build their own think tanks.

6. Japanese firms consolidate their market share by rescuing bankrupt financial houses and even make hostile takeover bids. They offer millions of dollars through university endowments to secure long-term access to well-trained bright graduates, more tailored to their needs.

are insufficient to block Japanese penetration of that segment. Japanese firms seek an entry segment with high growth potential, because high volume is central to their success due to the huge capacity at their disposal through the captive, protected home markets provided by the large domestic economic-interest groups, or *keiretsu* (for a more detailed discussion, see Wright and Pauli, 1988). An illustrative example of the structure of one of these large economic-interest groups is presented as Figure 5.

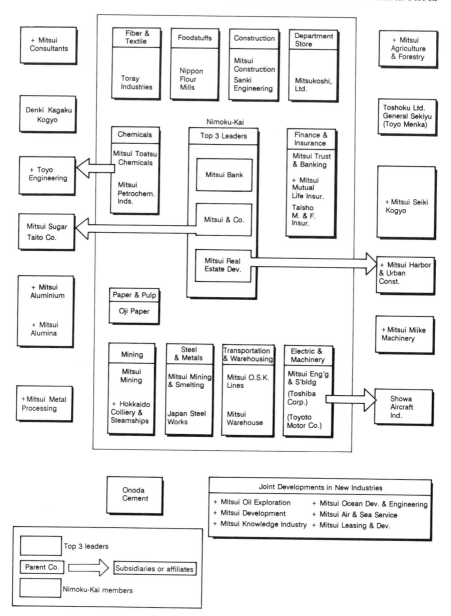

FIGURE 5   Mitsuigroup. SOURCE: Industrial Groupings in Japan. Revised edition, 1980/1981, Dodwell & Co.

Because price is used to capture a large market share quickly, it is important that the segment be highly price sensitive. They seek an area where there are few barriers to entry and where the entry risk to the Japanese is low. Finally, they seek a market segment that will open doors to the rest of the industry. These initial entry criteria are summarized in the first section of Table 2.

A simplified outline of the information technology market could identify seven subsectors (see Table 2, second section): semiconductors (chips), personal computers, minicomputers, mainframe computers, peripherals (e.g., printers and cables), communications equipment, and software.

For their initial entry into the information technology market, Japanese firms were not interested in the software segment, where they had performed poorly in the past. Also, they were not in a position to undercut the leadership of cash-rich European and U.S. competitors such as Lotus, Microsoft, Logica, and Cap-Gemini-Sogeti who were setting world standards. This segment still has not been tackled by Japanese firms. The mainframe market, which even in Japan was dominated until recently by IBM, did not appeal to them either as an entry segment, nor did the personal computer (PC) market led by Apple and IBM. These segments became the secondary targets.

Instead, the Japanese selected the semiconductor market as their first target in the information technology industry. Why? The U.S. competition was leading, but they were stand-alone companies, not integrated into large conglomerates as Japanese companies are in their *keiretsu* groups, with their large captive markets. U.S. companies had to pursue competitive strategies predicated on short-term profitability, to satisfy the cash-flow and dividend targets set by private investors and venture capitalists. The Japanese could aim simply for volume; they had their captive markets within the *keiretsu*, and whatever sales they made outside their group could be priced on the basis of marginal cost. Moreover, the basic semiconductor segment provided an ideal springboard to other, more sophisticated sectors of information technology.

The First Assault

Following this initial game plan, Japanese chipmakers successfully undercut the semiconductor market with quality products at very low prices. (Refer again to Table 1, step 1.) The economies of scale reached by the extra sales abroad served, as expected, to accelerate their learning curve in production and applications. In contrast to its U.S. counterpart, the Japanese semiconductor industry is not a merchant industry. Of the semiconductors produced by Matsushita, for example, 80–90 percent are used within its own group of companies. The remaining 10–20 percent sold overseas allows them to recoup extra research and development costs, pays for the marketing of the chips, and—perhaps most importantly—helps them to get a feel for the information technology market on an international basis.

Matsushita and the other large Japanese chipmakers were less constrained than their U.S. and European competitors by the huge capital investments needed to shift from 8- to 16- to 32- and 64-kilobit chips. There is almost always cash available in Japan's large industrial groups to expand production into areas of high demand. Moreover, the existence of guaranteed, stable internal markets within the groups lowers substantially the risks of large capital investments. The consumer sought low price and high product reliability, and the Japanese provided exactly that. They supplied a quality product in high volume at a low unit price. In the meantime the market kept on growing while a small number of competitors fought to maintain their share.

Branching Out

At the same time as the Japanese semiconductor manufacturers built up their market share in chips, forced prices down, and proved to clients that their product reliability was the best and their price the lowest, they were able to develop a parallel strategy of supplier contracts with original equipment manufacturers (OEMs) in other market segments, thus gaining valuable manufacturing expertise in related sectors (Table 1, step 2). Slowly but steadily Japanese firms began to penetrate not just the semiconductor market, but other related segments of the information technology market as well, such as microcomputers, peripherals, and mainframes.

Parallel to this, the Japanese improved their distribution system (Table 1, step 3). The initial presence of Japanese firms through independent distributors gave way to wholly owned subsidiaries for distribution and aftersales services. They then proceeded to consolidate their market position by replacing U.S. computers with Japanese models in government agencies at home and in selected countries abroad. Hitachi, for example, offered 50–60 percent discounts in Japan, Brazil, Spain, and Australia. They knew that once they had obtained a major market share of the government sector in those countries, the private sector would follow, purchasing their computers at higher prices. After all, one has to be concerned about dumping charges while selling to the private sector, but governments will seldom object to receiving computers for less than cost.

The Politics of Price

Soon after, in the wake of increasing pressure from Japanese competition and the apparent need for western companies to enter into coalitions with the Japanese, the cry arose from European and North American manufacturers: "The Japanese are dumping semiconductors." The Japanese invasion of the memory chip market between 1973 and 1975, and later on between 1981 and 1983 in the 64K memory market, triggered appeals for help, as the Japanese ability to underprice their competitors seemed unstoppable. On the basis of battles lost in the past, western companies such as IBM began to delegate production of their chips to the Japanese at the start of new projects, rather than fight a costly price war after large investments had been made in new production facilities.

Governments were pressured to block Japanese firms from the semiconductor market and to levy extra taxes on their products. To avoid the embarrassment of a dumping charge, the Japanese deftly resorted to "political engineering" (Table 1, step 4), working out an agreement on price and volume with the American private and public sectors. In the fall of 1985, the Reagan administration pursuaded Japan to sign a five-year agreement to stop dumping computer chips below "cost" (according to U.S. accounting standards) and to make its own semiconductor market more open to foreign manufacturers. However, the pact generated sharp controversy from the outset over its side effects, and U.S. charges of Japanese noncompliance led to the imposition of special retaliatory tariffs in the spring of 1987.

The outcome of the agreement was increased cost to the American consumer and increased profits for Japan. The chip deal between the United States and Japan resulted in an immediate average price rise of some 20 percent for semiconductors, with some chips even tripling in sales price. By forcing up chip prices in the U.S. market, the pact will handicap the competitive ability of other high-technology industries in the United States whose products contain semiconductors. Also, since the Japanese are no longer permitted to offer cutthroat prices on chips, they can increase their profit margins handsomely, thanks to the largesse of western policymakers.

Reaping the Profits

This additional profit from the consumer permits the Japanese to invest in chip factories in the United States (Table 1, step 5) in much the same way that they began investing in automobile plants in the United States after similar experiences in the automotive industry. There, the Japanese accepted a voluntary automobile-restraint agreement on their exports that allowed them—through the game of supply and demand—to increase their sales price by some $1,000 per car. After three years this extra income generated from the American consumer permitted the Japanese to begin building assembly plants in the United States. However, the higher value-added elements of car design and manufacture remain in Japan, as do the project-engineering and the production and process skills that underlie competitive success—the result of an implicit strategy of keeping the higher paying and higher value-added jobs in Japan. Then, for the sake of creating employment at whatever cost, the United States and Europe are happy to be delegated the lower value-added assembly functions. Nissan Motor's $575-million assembly plant in Smyrna, Tennessee, offers a case in point. Although the assembly itself takes place in the United States, the gear boxes and engine blocks—among the main value-added components—still come from Japan. According to the Japan Economics Institute, by the end of 1986 there were nearly 600 factories in the United States in which the Japanese owned a majority stake.

Aftermath

Today, the Japanese export more computer chips than do U.S. manufacturers, even though U.S. manufacturers held more than half the world market until 1981. As recently as 1982 the American share of the global market for microchips stood at 49 percent compared with Japan's 27 percent. By the end of 1986, Japan had taken the lead with 38 percent of chip sales versus 35 percent for the United States, according to In-Stat, an Arizona-based research firm. U.S. chipmakers do not expect much relief from competition through the agreement to limit dumping chips. They have already turned their attention to types of semiconductors that foreign rivals will find more difficult to copy. The U.S. chipmakers believe that their future lies in fabricating relatively small batches—thousands instead of millions—of customized chips for specialized uses. Companies such as National Semiconductor and Intel are moving away from commodity-like memory chips, to concentrate on microprocessors and other products that perform advanced functions. However, at least one company, Texas Instruments, has stubbornly refused to give up the race with the Japanese to make even more densely packed memory chips.

By expanding into products that use a lot of their semiconductors, Japanese firms have been able to take over world leadership of a wide variety of information technology products ranging from photocopiers, typewriters, and other peripheral equipment, to minicomputers; they now challenge even in telecommunications and mainframe markets (which will further enhance their software capabilities, because one-third of the mainframe market is systems software). Yet, U.S. and European firms were warned of this as early as December 1981 by a Canon executive in a *Business Week* interview: "We want to offer ALL products." If North American and European manufacturers had taken Japanese manufacturers seriously at that time, they could have prepared themselves better to meet the Japanese challenge in the information technology market.

THE JAPANESE STRATEGY IN FINANCIAL SERVICES

Japanese firms are now applying an incremental strategy to the financial services industry as well, and the path is remarkably parallel to that followed so successfully in information technology. Again there is no master plan, but in the end it becomes a clear and well-balanced strategy to get to, and stay at, the top of the list of the world leaders of financial services. The scenario outlined below is already well under way.

Targeting Governments

Let us, for this discussion, simplify financial services into seven categories (refer again to Table 2): government lending, foreign exchange, underwriting, stock exchange operations, trust banking, swaps, and mergers and acquisitions.

TABLE 2   Market Entry Considerations

Characteristics of entry segment most attractive to the Japanese:
1. A segment where the financial resources of the competition will not permit major resistance and counterattack to the Japanese assault on that segment
2. A segment characterized by double-digit growth (i.e., enormous volume)
3. A highly price-sensitive segment
4. A segment where entry risk for Japanese is low, and with few barriers to entry
5. A segment that provides an entry ticket to other segments

Segments of the information technology and financial services markets:

Information Technology	Financial Services
1. Semiconductors (chips)	1. Governmental lending
2. Personal computers	2. Foreign exchange
Peripherals	Underwriting
Minicomputers	Stock exchange operations
Mainframe computers	Trust banking
Communications equipment	Swaps
3. Software	3. Mergers and acquisitions

Key characteristics of the semiconductor and government lending markets:

Semiconductors	Government lending
1. Small number of competitors	1. Few lenders have enough cash to supply demand
2. Standardized high unit volume	2. Huge government appetite requiring a similar approach
3. Price sensitive	3. Price sensitive
4. Low risk, if only surplus products are sold to outsiders	4. Low risk because governments generally do not go bankrupt
5. Major entry ticket to other information technology products	5. Important for creation of political goodwill

Key factors for success in semiconductors and government lending:

Semiconductors	Government lending
1. Product reliability	1. Placement capacity
2. Low price	2. Low margins over cost of funds

Japanese firms have not tackled the financial services market through the mergers and acquisitions business—this will be the last segment for them to attempt—or through trust banking, because they are noted for their inexperience in these sectors, both at home and abroad. Recognizing their limited know-how, they seek cooperative efforts in those areas, with the full support of their government.

Rather, Japan's initial entry into the market for financial services has been mainly through government lending. (Refer back to the flowchart of Table 1 for the remainder of this discussion.) The appetite of governments for extra funds, their somewhat standardized approach to loan deals, and the high volume and price sensitivity of government lending, all provided the Japanese with sufficient justification to conclude that this was the place to start. Few competitors can satiate the cash needs of governments. In this highly price-sensitive market, the Japanese offered loans priced often only 1/16 of one percent over their cost of funds, quickly making them the favorite of governments the world over. The county of Los Angeles, for example, turned to the Japanese for a $250-million loan, as did the New York Job Development Agency for a $190-million loan arranged by Sumitomo. Japanese institutions always offer slightly better rates than any other banks, go for minuscule margins, and provide consumer satisfaction in much the same way they satisfied the person who bought his or her first Toyota at a discount price and got "hooked" on a quality product with a long-term warranty.

Learning from Others

With government lending well established for Japanese institutions, they have begun to join with foreign financial institutions in other segments of the financial services market, such as foreign exchange, stock exchange operations, and trust banking. They are actively acquiring expertise in much the same way that the information technology industry got a feel for the market through the OEM agreements. Meanwhile, behind the scenes, the Japanese government helps smooth the way wherever possible.

For example, when trust banking was deregulated in Japan, the Japanese government required foreign entrants "to employ staff knowledgeable of Japanese trust business, particularly pension fund business." Yet how could any foreign financial institution have such expertise when the market had always been closed to them? With the traditional lifetime employment system and company loyalty in Japan, it was almost impossible for a foreign bank to hire away local experts. The only way to meet the requirement was to join forces with a Japanese trust bank: the foreigners offering their sophisticated know-how in return for an opportunity to participate in the market. Citibank teamed up with Yasuda Trust and Banking; Chase Manhattan and Manufacturers Hanover, with Daiwa Bank; J. Henry Schroeder Bank and

Trust was bought by the Industrial Bank of Japan. The likely results were summed up succinctly by an American banker in Tokyo: ''None of the foreign trust banks here are ever going to make much money, but the Japanese are going to learn everything there is to know about trust banking through the joint ventures.''

Branching Out

At the same time, Japan's financial houses began building an impressive international infrastructure. In 1975 the Japanese banks had only 225 branches abroad, but by 1985 this had risen to 716, and these branches are well connected with their headquarters to secure efficient management of the operations worldwide. Sanwa, just to mention one, invested the equivalent of $8.4 million over the past two years in the Sanwa Overseas Banking Automation System. Here, once again, the intimate ''family'' relationships among companies within Japan's huge industrial groups play a central role. Financial institutions in Japan have immediate, proprietary access to state-of-the-art computer and communications technology from other industrial companies of their same group, rather than having to seek and acquire it at arm's length on an open market, as in the United States. This provides the Japanese financial houses a substantial edge in setting up an advanced technology infrastructure.

Japanese financial institutions have also embarked on a dramatic series of acquisitions abroad to obtain new expertise and to gain market share. Their aggressive acquisition strategy has so far been limited in Europe to one particular case: the purchase of 52 percent of Banca Gottardo of Switzerland by the most profitable of the Japanese banks, Sumitomo. However, similar acquisitions are bound to occur as opportunities arise, particularly following the inevitable shakeout resulting from the recent ''big bang'' in London.

Japanese moves to buy expertise and market share are much more evident in the United States. It is worthwhile enumerating some recent examples, because they provide insight into the bold strategy of the Japanese—steps individually conceived and executed, although they all seem to form part of a major coordinated assault on the market:

1983: $475-million purchase of the Chicago-based commercial financing company Walter E. Heller by Fuji Bank;
1985: $520-million purchase of the Continental Illinois leasing subsidiary by Sanwa Bank;
1986: $500-million purchase of 12.5 percent of Goldman Sachs stock by Sumitomo Bank;
1986: $250-million purchase of the Bank of California by Mitsubishi Bank, which had previously acquired BanCal Tristate for $282 million;
1986: Bank of Tokyo purchase of California First Bank;

1987: Purchase by Nippon Life, the largest life insurance company in Japan, of a 13 percent stake in Shearson Lehman Brothers, the securities operation of American Express, for $538 million; and

1988: Nomura Securities' acquisition of a 20 percent stake in Wasserstein Perella, a leading mergers and acquisitions firm, for $100 million.

The approach has been well targeted. Today, half of the 12 largest banks in California are Japanese owned.

Seeking Greener Pastures

Financial services closely related to banking and securities are next on the target lists of the Japanese. JCB, the dominant Japanese credit card, was launched in 1961 by Sanwa Bank when there were only 100,000 Japanese travelers abroad. It benefited for years by collaboration with American Express. Today, with some four million Japanese traveling, JCB has abandoned the cooperative relationship with American Express and plans to compete with them directly. JCB is now represented in 30 countries with more than 25,000 outlets; within two years their infrastructure will spread over 60 countries and some 50,000 outlets.

Leasing is still another area where the Japanese are drawing on expertise from others to expand abroad. Sanwa Bank led the way by enticing Dresdner Bank to join them in a joint venture with the Bank of China, for a major leasing firm based in Beijing.

Japan's securities houses are expanding and diversifying even more aggressively than the banks. They have successfully obtained commercial banking licenses in Australia, the United Kingdom, and Luxembourg; primary dealer status on the New York exchanges; and broad penetration of many smaller financial markets such as Amsterdam, Hong Kong, and Singapore. They were able to gain entry through well-crafted political moves and bilateral agreements, whereby the Japanese government offered certain inducements to foreign governments in return for major concessions to the Japanese houses abroad.

A good example is the trade-off of two seats for British brokers on the Tokyo Stock Exchange, in return for a seat on the London Stock Exchange and a commercial banking license in London for Nomura Securities Company. It is interesting to observe Japan's Ministry of Finance working so diligently to promote activities of Japanese financial houses abroad which would be prohibited to those same firms (as well as to foreign firms) by Ministry of Finance regulations at home.

Future Steps

Under the most likely scenario, we may expect soon to hear a crescendo of charges that the Japanese are "dumping" financial services. Indeed, the chairman of Barclays Bank recently leveled an open accusation against them for doing just that. The Japanese are currently offering loans at only 1/16 percent over cost, and their fees for funds management are only 0.55 percent when the market standard is 1 percent. Pressure is mounting on governments in North America and Europe to negotiate "voluntary" restraint agreements with the Japanese in financial services, as was done in semiconductors and in textiles, steel, and automobiles before that. Also, just as with those other industries, protectionism in financial services will allow Japanese firms to raise their profit margins and to use the additional capital for extended joint ventures and straightforward acquisitions abroad, thus strengthening and further consolidating their global position in financial services.

The expansion of Japanese global financial markets threatens North American and European financial institutions. As long as governments, the consumer, and the sector concerned know that they have to face up to the Termite Strategy, it is possible to design effective counterstrategies. Unless U.S. and European policymakers understand this strategy, they will design the wrong policies, ask for protection that will not help, and seek cooperative arrangements that will result in the draining of national expertise.

REASONS FOR CONCERN

At first glance, it is easy for Americans and Europeans to shrug off or even to welcome the new Japanese competition in financial services. After all, if they are willing to supply us with large amounts of capital at cut-rate prices, why complain? Who isn't glad to get money cheaper? However, beyond that immediate boon to the consumer of funds, the potential consequences of Japanese dominance of global financial services are real and substantial.

A first consequence is the obvious—although often overlooked—fact that the piper must eventually be paid. As the United States and other industrialized countries of the West move away from their traditional capital self-sufficiency or even capital export positions toward mounting external indebtedness, so do their obligations for future payments of interest, dividends, principal, and service fees to foreign creditors mount relentlessly. Unless these payment obligations are matched by corresponding increases in real productivity—a most unlikely prospect—then more and more of our productive energy and that of future generations will be devoted to servicing external financial obligations.

Moreover, as western economies become increasingly dependent on sustained financial support from outside sources, so they become more vulnerable to economic and political influences from abroad. Already, for example, the United States has largely lost its ability to manage domestic interest rates independently, for fear that lower rates would trigger a destabilizing outflow

of Japanese investment capital. It is no exaggeration to say that the United
States is becoming an "underdeveloping" part of the globe, slipping toward
a position of financial dependence not unlike that suffered today by many
lesser developed countries. Japan's influence goes well beyond government
bonds and corporate securities. *Business Week*, in a recent survey, concluded
that "Japanese companies are spending heavily to shape the way Americans
view them. . . . The Japanese are also wielding political power from the
grass roots to the top echelons of Washington" (*Business Week*, 1988).

Perhaps even more serious is the potential long-term threat to U.S. economic vitality and entrepreneurial spirit. As U.S. domestic financial institutions fall increasingly under direct foreign control, which they will if current
trends continue, the highest level, highest value-added financial skills—
planning, strategy formulation, engineering complex financial packages, even
finance-related technological developments—will be increasingly centered
in Tokyo or Osaka.

Awareness and Complacency

The worst enemy of the western economies is complacency. It is too easy
to shrug off the Japanese financial challenge with the same platitudes used
a decade ago as Japanese firms become increasingly important in manufacturing industries, from watches and cameras to cars and information technology: "It's a passing phenomenon"; "the Japanese can't really innovate";
"Japanese business is so culture-bound that it will quickly flounder outside
of its homeland"; "U.S. and European managers are innovative enough to
keep ahead of them."

The growing Japanese presence in financial services may bring short-term
economic and political benefits, but it also can have far-reaching implications
in the longer run on national trade balances, the quality of employment, and
national innovative abilities. Japanese firms do not often publicize their
strategies and their competitive successes as western firms are prone to do.
They prefer to keep a low profile; thus their expansion moves go unnoticed.
However, it is time for western business and government leaders alike to
open their eyes to the challenge of Japan's second wave, to acknowledge
that there is a major new competitive challenge, and to reshape their competitive strategies accordingly.

DESIGNING EFFECTIVE RESPONSES

Our discussion of the Japanese strategy has identified some key organizational technologies and competitive elements underlying the success of
Japan's financial services firms. Among the most significant are: (1) access
to very large amounts of capital at very low cost; (2) stable, well-educated,

highly-motivated work force; (3) an ability of corporate managers to plan and target resources effectively; (4) a long-term profit horizon permitting Japanese firms to focus on long-term price-cutting and other market-share competitive strategies, rather than short-term profit; (5) large, stable internal markets and customers, through the *keiretsu* group relations; (6) direct access to state-of-the-art electronic and communications technologies, again through direct, proprietary access to other firms in their keiretsu groups; (7) government policies that both protect the home market and actively promote the position of Japanese financial institutions abroad.

Some of these advantages clearly stem from cultural factors and institutional relationships that cannot be replicated readily by others. Nevertheless, U.S. and European financial institutions still hold a firm lead over the Japanese in many high-value-added areas of financial services. With careful responses on the part of both the financial institutions and governments, there is much that can be done to maintain U.S. and European leads in those areas and strengthen the general competitive position of western institutions.

Creating a Competitive Edge

Once western financial houses shed their complacency and accept that they indeed face a crisis, their first critical need is to define their strengths: the areas where their productive resources can be focused most profitably for the future. One of the great strengths of Japanese managers is their ability to target their competitive energies, focusing on specific niches where their potential strengths are greatest or where their competition is weakest. There are many high-value-added financial activities in which western financial institutions still have a clear-cut edge over the Japanese: trust banking, portfolio management, venture capital, mergers and acquisitions, swap arrangements, and engineering complex financial packages, to name a few. Given limited resources, North American and European managers should direct their energies toward these and other new niches that will certainly arise in the highly fluid environment of international finance.

Keeping a step ahead of the competition will mean constantly scanning, innovating, and targeting new directions. U.S. and European financial houses certainly are creative enough to do this. After all, nearly all the significant innovations in financial services to date have originated outside Japan. The real challenge will be to shift quickly to the new, high-value-added activities as they open up, even at the cost of relinquishing portions of our more traditional markets. This will be difficult, as it will require deemphasizing or even abandoning some familiar activities and markets in which both institutions and individuals have strong vested interests. Nevertheless, the only alternative to focused change is to continue serving traditional markets behind increasingly restrictive protectionist legislation, raising the cost to consumers and ultimately eroding competitive abilities even further.

Although better targeting will help, there is also more that managers can do to attend to the needs of their customers. Western firms cannot realistically hope to compete head-on with the Japanese giants in price competition: they lack access to the supplies of low-cost funds of the Japanese; and the demands of American shareholders and creditors for short-term profit performance preclude most financial houses from sustaining for substantial periods the low earnings that price-cutting usually implies. To compete with Japanese institutions, western financial firms will need to focus more on relationship services aimed at client loyalty, rather than on transactional services susceptible to price competition. What this boils down to is a more market-oriented approach to winning and holding the loyalty of customers, be they individuals or corporations. If financial institutions provide optimal service within a long-term framework of confidence, most customers will not turn their backs for a mere one-half or one-quarter percent.

More can be done also in the internal management of U.S. and European financial institutions, particularly in human resource management. One of the main strengths of Japanese competitors is their low labor turnover and the extremely strong loyalty of their employees. Cultural differences clearly preclude the same degree of company devotion in most western societies. There is, nevertheless, a great deal that can be improved in recruiting, training, job security, and employee participation that could help U.S. and other western companies improve their long-term productivity. One of the greatest competitive strengths of U.S. and European institutions is a proven level of creativity and entrepreneurship. The ability to motivate personnel through new approaches, such as profit-sharing schemes that give larger numbers of employees a feeling of partnership and shared accomplishment, can be a key determinant of success in countering the challenge of the second wave.

Building and maintaining an effective competitive edge will require also a significant reorienting of western shareholders and investors. The relentless demands placed on western financial institutions for short-term profit performance constitute one of the most serious handicaps in facing the Japanese. Whereas the best route to meeting the Japanese challenge is clearly not to try to match their price competition head-on, western financial services companies could nevertheless compete much more effectively if they were able to develop strategies based on long-term growth and market-share considerations, just as the Japanese do, thus enhancing their ability to engage the Japanese on their own terms where appropriate.

Organizational Linkages

At the same time western financial houses face growing competitive pressure to specialize their focus into niches of comparative advantage, they face seemingly contradictory demands for greater diversification of the services they offer. It is clear that consumers of services increasingly seek "one-stop shopping,'' i.e., financial services centers capable of providing a broad range of services, from straight financial loans and deposits to equity trading, insurance policies, currency swaps, and travel services. Japanese financial service institutions are diversifying rapidly to meet this demand, achieving economies of scope as well as of scale. Few individual western companies are yet equipped to provide such a broad range of services efficiently, regardless of size.

One appropriate response is to form linkages with other home-based financial institutions offering complementary high-value services. This need not be done through formal mergers or acquisitions. Indeed, the recent spate of mergers and acquisitions in western financial services industry—based more on leveraging balance sheets with undervalued assets than on any market strategy guided by complementarity—has sapped the vitality of many companies by tying up huge amounts of capital in nonproductive stock purchases, as well as by demoralizing the organizations involved and causing a flight of personnel. What is needed, instead, are informal alliances among firms offering complementary financial services—i.e., banks, insurance companies, brokerage houses, foreign-exchange specialists—all linking up contractually to provide their own specialized services in combination with others. Such alliances will facilitate integration of financial services and the fusing of technology, thus improving both efficiency of the companies and services for their customers. The resulting increases in scale, scope, and efficiency can provide a substantial counterweight to the Japanese onslaught.

A related type of alliance is with other cash-based service companies such as retail stores, aimed at securing access for financial services to new mass-distribution and even electronic-distribution channels. Such alliances help bring organizational economies of scale to the firms involved, increasing their productivity and better serving the convenience of their customers. Again, we have only to look at the highly diversified *keiretsu* groups in Japan to see how successfully diversification through collaboration among complementary, like-minded companies can enhance international competitiveness.

Even more far-reaching may be cooperative working relationships among potential competitors of different nationalities, including the Japanese themselves. In other words, one approach is to combine U.S. financial skills and marketing expertise with Japanese access to capital and placement channels, to provide the most efficient possible symbiotic combination of comparative advantages. At first glance, the idea of forming alliances with competitors may seem preposterous, but that is exactly what is happening in some manufacturing sectors, most notably automobiles. Who would have thought a few years ago that we would ever see the networks of global working alliances that exist today between General Motors and Toyota, for example, or among Nissan, Volkswagen, and Alfa Romeo? The important point here is that it is possible to achieve mutually productive working relationships on an international scale on the basis of cooperative agreements short of formal mergers or acquisitions, thus helping to maintain the independence and integrity of each of the participating companies.

A striking example of this collaborative potential between Japanese and westerners in financial services is American Express Company's recent sale of a 13 percent stake in its Shearson Lehman Brothers securities operation for $538 million to Nippon Life Insurance Company (*Financial Times*, 1986 and 1987). This follows Sumitomo Bank's earlier purchase of a 12.5 percent investment in Goldman Sachs for $500 million (although the latter deal excluded a direct equity link). The latest move brings together some of the most massive players in the financial services industry: American Express is one of the world's largest international financial services conglomerates; Nippon Life is the largest of Japan's huge insurance companies; and Shearson is the third-largest Wall Street investment bank. In a smaller but equally significant move, giant Nomura Securities Company paid $100 million for a 20 percent stake in Wasserstein Perella, a leading mergers and acquisitions specialist, moving the Japanese for the first time into the highly sophisticated mergers and acquisitions business (*The Economist*, 1988).

The trick in such cross-national collaborative arrangements is, of course, to ensure that the long-term gains to the two sides are really balanced. Too often in the past, such arrangements appear to have helped the Japanese gain significant competitive advantages by giving them access to valuable skills, while offering their U.S. and European partners little in return except short-term infusions of capital or potential market access of dubious value. A recent *Harvard Business Review* article (Reich and Mankin, 1986) on collaborative arrangements with Japan concludes that

The big competitive gains come from learning about . . . processes—and the result of the new multinational joint ventures is the transfer of learning from the United States to Japan.

There are indications that westerners are beginning to scrutinize much more carefully the trade-offs involved in collaborative arrangements with the Japanese. A major objective of Sumitomo's acquisition of 12.5 percent of Goldman Sachs—to acquire skills in investment banking—was thwarted, for example, when U.S. officials refused to permit Sumitomo to send trainees to Goldman's New York offices, on the grounds that such activity would violate the statutory separation of commercial and investment banking.

Government Policies

By identifying and pursuing new profit opportunities, by attending better to the needs of both customers and employees, and by forming effective alliances with other services companies both at home and abroad, U.S. and European financial services firms can do much more to develop client relationships that are strong and loyal. Yet to sustain any real competitive edge, the private sector needs one other vital ingredient that Japan's banks and securities companies can safely rely on: a sympathetic and supportive government policy environment.

This emphatically does not mean protectionism. It is clear from earlier attempts to cope with Japanese manufacturing competition that protectionism is not a viable long-run solution. Artificial barriers imposed by the United States against the import of low-priced Japanese semiconductors will be ultimately counterproductive to U.S. competitiveness in the higher value-added end of the information technology market, where U.S. firms still remain competitive. Similarly, the United States, which remains at the forefront of many high-value financial functions, has far more to lose than to gain from restricting the international flow of financial services.

Instead, western governments should continue deregulating financial services at home, facilitating greater efficiency in their domestic firms. Recent moves in this direction by Canada and the European Community are encouraging; but the United States lags far behind. Regulations limiting branch banking, and most particularly provisions of the Glass-Steagall Act, which segment the various types of financial services activities, handicap financial houses from achieving the scale and scope of activities needed in today's global competition. They are counterproductive and should be removed. western governments—and again, Americans in particular—will almost certainly need also to relax current antimonopoly legislation to encourage the kinds of interinstitutional cooperation needed to square off against the concentrated strength of the Japanese.

At the same time, the U.S. government must continue striving to bring about the "level playing field" needed if U.S. institutions are to compete abroad on an equal footing with foreign competitors. This means, above all, continued relentless pressure on the Japanese government to allow foreign firms the same access to Japanese financial markets that Japanese houses enjoy abroad. Significant steps have been taken to open the Japanese market, but major differences of opinion remain as to what reciprocity actually means. U.S. financial services firms claim—correctly—that Citicorp and Merrill Lynch are not permitted to do in Tokyo what Dai-Ichi Kangyo and Nomura can do in New York. The Japanese politely respond that they treat a Merrill Lynch in Japan exactly as they treat a Nomura: according to the Japanese book of rules.

Finally, the U.S. government needs to do more to make accessible to U.S. firms the ultimate competitive advantage: low-cost capital. Current government fiscal policies, especially massive government deficit spending, serve to drive up interest rates in the United States, increasing the cost of capital to U.S. financial services houses and undermining their global competitiveness. Monetary and fiscal policies leading to low inflation and low interest rates will ultimately be more helpful than any amount of protection or subsidies.

CONCLUSION

Meeting the challenge of Japan's second wave will require bold new approaches by financial houses and governments alike, and most important, it will require an extraordinary willingness on the part of private and public sectors to work together. There is no clear-cut formula or set of steps that westerners can take to counter the Japanese challenge in financial services. However, by being aware of the Termite Strategy—of what is happening, where the Japanese are likely to go, and how they intend to get there—U.S. and European institutions can better prepare themselves to shape appropriate competitive strategies for the future, not only with respect to the immediate financial threat, but also with respect to the broader emerging challenges of international trade in services.

REFERENCES

Business Week. 1988. Japan's influence in America. (July 11):64–75.
Echo de la Bourse. 1987. The brokers (April 23).
The Economist. 1986a. Richer than you. (October 25):13–14.
The Economist. 1986b. Topsy-turvy. (March 22):15.
The Economist. 1988. Big wheels, small deals (July 30):78.
Euromoney. 1985. Who's doing business. (August):151.
Financial Times. 1986. Nippon Life to pay $350m for Shearson stake (March 20).
Financial Times. 1987. American Express hunts for a global advantage (March 25).
Hector, Gary. 1986. The Japanese want to be your banker. Fortune (October 27):97.
Reich, Robert B., and Eric D. Mankin. 1986. Joint ventures with Japan give away our future. Harvard Business Review (March-April):78.
Time. 1986. Money masters from the East. (August 11):31.
Time. 1988. Yen power goes global.(August 8):20–23.
Wright, Richard W., and Gunter A. Pauli. 1988. The Second Wave: Japan's Global Assault on Financial Services. New York: St. Martin's Press.

Key Policy Issues Posed by Services

JAMES BRIAN QUINN AND THOMAS L. DOORLEY

Other chapters in this volume document the scale and importance of services in the U.S. economy and the role of technology in improving the productivity of, and the value added by, the services sector. Given the dominance of services activities in the U.S. economy, policies supporting technology in services should be an important component of the ongoing debate about the government's proper role in supporting technology for both public and selected private purposes, particularly policies for improving U.S. industrial competitiveness. This chapter focuses on a few themes of particular relevance to the "national competitiveness" debate. Two issues are of special importance:

1. The effectiveness of U.S. services industries and U.S. manufacturing are mutually intertwined. Attempting to strengthen one without strengthening the other would be a misguided policy. Unfortunately, to date there has been a tendency to focus competitiveness policy discussions almost exclusively on manufacturing issues. Effective services activities actually create new markets for manufactured goods, result in lower costs for manufacturers, and are central to increasing the value added by manufactured products. Similarly, manufacturers are important suppliers, customers, and innovators for services activities.

Not only are services and manufacturing mutually reinforcing, the same policy approaches that stimulate or retard one sector will generally affect the other in similar ways. However, the impact of policy actions will tend to be proportionally greater on services because of the larger scale of the services industries, which provide 75 percent of all employment and 71 percent of all U.S. gross national product (GNP). Critical factors affecting both include the high relative cost of U.S. capital (which may benefit from a general leveling of such costs worldwide induced by the global integration of capital markets), the skill and literacy levels of the work force, and the development and application of frontier technologies (particularly those related to communications, health care, information handling, and public transport).

2. Technologies in the services sector are restructuring many manufacturing and services industries—as well as the entire U.S. economy and its international trade patterns—in ways that make past "industry-focused" approaches to regulation or trade relations even more inappropriate. The dynamics of technology, combined with deregulation, have broken down barriers among industries such as transportation, communications, finance, distribution, education, and health care, and have created a degree of cross-industry interaction and competition that calls for new regulatory philosophies and institutions across a wide spectrum.

Later sections of this chapter attempt to coalesce many of the significant policy themes developed by other chapters in this volume and to make specific recommendations concerning these major focal points.

BACKGROUND ASSUMPTIONS

Services are central to employment, economic growth, and quality of life in an advanced industrial nation such as the United States. However, unlike manufacturing, services have lacked a coordinated advocacy in policy circles. The chapters in this volume strongly suggest the following as essential background assumptions for a balanced U.S. policy discussion:

• This nation's strong services sector is a natural and desirable outgrowth of a highly productive industrial economy and the sophisticated application of technology to services activities. Services growth has not led to a decline in the overall manufacturing base; rather it has both created new markets for manufactured goods and supported increased manufacturing competitiveness. Despite the serious concerns expressed about declines in specific manufacturing industries in the United States (and in other major developed countries), total employment in U.S. manufacturing has fluctuated around a zero trend line for a long while, and real manufacturing output and value added have continued to grow (see Figures 1–3 in Quinn, this volume). On the other hand, services have been—and will continue to be—the nation's driving economic force in both arenas in the near future. A large majority of the most successful new ventures of the last two decades (1960–1980) were in services (see Table 1), and many of the most successful new manufacturing ventures (such as Apple, DEC, and Wang) sold products largely to the services sector (see Roach, this volume). Productivity statistics suggest that there is no inherent reason why large services industries cannot improve their productivity as rapidly and as much as manufacturing. In many services industries, technology can also be effectively leveraged to create competitive advantage and higher margins. In other services industries, productivity and quality improvements must be rapidly passed through to

TABLE 1   Examples of Most Successful New Enterprises, 1960s and 1970s

Service, 56%	Service-Product, 9%	Product, 35%
H & R Block	Mary Kay Cosmetics	Intel
People Express	Computervision	Honda
Hospital Corp. of America	Pizza Hut	Sony
Microsoft	Mrs. Fields' Cookies	Apple
Hambrecht & Quist	Damon	Nike
LucasFilms, MCI	McDonald's	DEC
Wal-Mart		Tandem
EDS, Tandy		Apollo
Federal Express		Wang
Holiday Inns		

SOURCE: Compiled from various publications' listings of most successful new companies.

customers, adding great complexity to the interpretation of output and productivity measures.

• Services enterprises, such as banks, communications companies, airlines, and health care providers, are now among the most significant initiators, users, and managers of technological systems. Various chapters in this volume and its companion volume, *Managing Innovation: Cases from the Services Industries* (Guile and Quinn, 1988), illustrate how technologies developed or implemented by such enterprises are revolutionizing long-established economic and trade relationships within and among nations, thus raising profound new policy issues for both business and government.

• Services are an increasingly important component of international commerce. The fact that countries have begun to recognize the importance of services in economic development (Faulhaber et al., 1986; Shelp, 1986) and the focus on services in the Uruguay round of the General Agreement on Tariffs and Trade (GATT) are indicators of this trend. As Wright and Pauli make clear in their chapter, foreign governments' targeting of specific services for trade development is becoming a high-profile concern, particularly as the services companies of those nations begin to take over attractive U.S. services enterprises and aggressively expand their influence in U.S. and world services markets.

Unfortunately, greater understanding of the importance and potentials of the services industries is often obfuscated by mistaken attitudes about services, limitations in available data, and the misleading measures used for important policy determinations.[1]

In particular, difficulties in measuring services productivity (especially the amount and quality of services outputs) and structural problems in utilizing technology within some services activities (particularly personal and professional services) often obscure the importance of technology in services—and of services in the economy (Mark, 1982, 1986). Serious funding and administrative support are needed to modernize and upgrade both domestic and international trade data concerning services.

THE SERVICES-MANUFACTURING INTERFACE

The services- and goods-producing sectors are so intertwined that it is counterproductive to think of policy mechanisms for one without carefully considering the impact on the other. The interactions between services and manufacturing cover a broad, and often unrecognized, spectrum of activities and profoundly affect the performance and competitiveness of U.S. manufacturing enterprises (see Figure 1). Although a chapter in the companion to this volume (Guile and Quinn, 1988) makes this point in greater detail, it is important enough to highlight for policy purposes here. Many have noted that U.S. services industries are often dependent on relationships with manufactured goods, i.e., their businesses exist largely by providing transportation, finance, advertising, repair, distribution, or communications supporting transfers of manufactured goods. Interestingly, however, many of these services would still be provided in the United States, regardless of where the product was manufactured. Although some design support functions and supplier interlinkages could move overseas if manufacturing were not performed here, even these functions are increasingly being performed remote from manufacturing sites. Because of technological economies of scale and technological innovations, services industries—both upstream and downstream from manufacturing—appear to be actually increasing their leverage versus manufacturing.

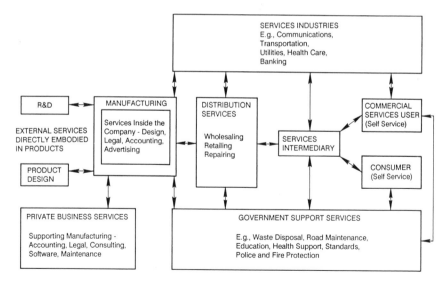

FIGURE 1   Some mutual interactions among manufacturing and services activities. Double-headed arrows indicate that each party benefits from the presence of the other in trade.

Less recognized is the dependency of manufacturing on services. Some 70–85 percent of all high-technology (information) products are sold to the services sector. In addition, manufacturing success today often requires more rapid feedback from the marketplace, better customized products, and more accurate delivery in shorter cycle times—all of which are dependent on downstream services integration. Proper integration of these technologies throughout manufacturing and distribution could significantly increase the number and range of goods that it pays to produce in the United States, rather than overseas.

• "Quick response" reordering systems can allow U.S. textile mills significant competitive advantages. They make it possible to deliver fabric in one-third the time it takes from Taiwan, and computer-assisted design and manufacturing links between cutters on Seventh Avenue and southeastern mills can halve the time it takes from design to goods delivery (*Business Month,* 1987). Well-managed retailers and distributors increasingly know what customers want better than manufacturers possibly can. Many are using their market knowledge and electronic point-of-sale (EPOS) data systems to actively participate in designing products and guiding manufacturers' sales strategy through the improved marketing data they provide.

U.S.-based manufacturers able to link flexible production systems directly to their customers' market intelligence networks should have sustainable competitive advantages (both in timing and in transportation costs) over foreign producers. As foreign firms invest in U.S.-based manufacturing capacity to get closer to the country's huge, increasingly customized marketplace, this is leading to a substantial "remanufacturing" of the United States. The Japanese automobile companies' recent moves into the United States provide interesting examples of the potential impacts on both producing and supplier industries.

In addition, services technologies offer a rich new array of channels through which manufacturers can reach specialized segments of their markets. Electronic home shopping and interactive video terminals located in banks, airports, hotels, airplanes, and shopping malls allow manufacturers to make contact with whole new groupings of customers in psychological situations in which they are likely to buy.

Manufacturers Becoming Services Providers

Increasingly, the profitability of manufacturers depends upon their use of services technologies and their extensions of these technologies as "products" for exploitation.

The competitive positions of large companies are now largely determined by their capacities to manage information worldwide—about suppliers, new technologies, exchange rates, swap potentials, or the changing political or market sensitivities in key countries. For example, with crude oil resources primarily in the hands of sovereign nations, Exxon's profits depend ever more on its ability to find, track, deploy, trade, transport, finance, and distribute energy efficiently. All of these activities are "services" and are technology driven. In addition to such logistics activities (especially in world-wide sourcing), General Motors (through its Acceptance Corporation, GMAC) has found financial services to be an indispensable competitive weapon in the marketplace and a source of over half of its profits in recent years.

In many companies such as IBM, services technologies (e.g., software) have always been a key to success, often being "bundled" integrally into product and rental strategies. Now, with the hardware aspects of many electronic products becoming extremely low cost and competitive, these "manufacturing companies" are increasingly shifting their focus toward software, networks, and communications linkages (services) as their bases for improving value added and profits. Others, such as Caterpillar Company, have found new growth through providing logistics and transportation services (utilizing their own extensive world networks) for other manufacturers interested in efficient warehousing or parts delivery worldwide.

Lowering Internal Manufacturing Costs

Many aspects of a manufacturer's cost competitiveness depend intimately on services activities, either within the firm or purchased from vendors. Greater efficiencies in communications, transportation, financing, distribution, health care, or waste handling (services industries) can markedly affect a manufacturer's direct costs. To the extent that these services are more efficiently provided, they lower living costs for workers and improve the quality of life they can enjoy at any given wage level.

In addition to the 75 percent of U.S. employment directly in the services industries, within manufacturing businesses an astonishingly large proportion (estimated to average some 75 percent) of all costs—and a much higher percentage of value added—are generally due to services activities (Office of the U.S. Trade Representative, 1983; Vollmann, 1986). Aggressively managing services activities within manufacturing enterprises can provide a major attack point for improving competitiveness in the future. As advanced technologies provide new economies of scale or scope to specialized services providers, manufacturers are increasingly finding that they can substantially improve their costs and effectiveness by "outsourcing" staff services such as accounting, personnel, legal, marketing, and even research and design functions.

Services and International Manufacturing Operations

One of the areas in which services technologies affect manufacturing most significantly is international operations. Telecommunications, air transport, and surface cargo handling technologies have forced virtually all manufacturers to consider their supply sources, markets, and competition on a worldwide scale or lose their competitive position. Although the figures do not show up in merchandise trade balance statistics, the greatest impact of U.S. manufacturing technology on world markets is probably through the operations of multinational companies' inside host countries (Sauvant, 1986). About one-fifth of the total capital invested in U.S. manufacturing firms is in facilities outside the United States, with a similar proportion of output produced there. Some of the largest continuing favorable net balance of trade accounts for the United States have been the profits, royalties, and intercorporate sales benefits remitted by these multinationals to the United States (see Table 2).

Effective coordination of the international operations of large manufacturing enterprises and of many much smaller companies depends heavily on services technologies and efficiencies. Also, economies of scale in international operations are very often due to the corporation's services capabilities (i.e., technology transfer, marketing skills, financial services, or logistics) rather than its plant scale economies. A significant component of a multinational company's competitive edge comes from its capacity to handle cross-border data and services flows. Consequently, maintaining the freedom of these flows is a very sensitive and critical point in maintaining the international competitiveness of U.S. manufacturing enterprises. International manufacturing operations and services technology management are inseparable for producers seeking competitive advantages in today's global marketplace.

Manufacturing's Changing Strategic Environment

As has been noted, perhaps the most important structural change in international manufacturing competition stems from the continuing integration (through electronics) of the world's financial centers into a single world financial marketplace. World financial flows have already become largely disconnected from trade flows (Bell and Kettell, 1983).[2] Differences in national economic policies can disturb interest rates only slightly but still call

TABLE 2   Manufacturing-Related Trade Balances ($ Billion)

Category	1970	1975	1982	1985	Preliminary 1987
Royalties and fees, net	2.1	3.8	4.6	5.3	6.8
Direct investments, net	2.6	14.4	18.2	26.6	35.3
Merchandise, net[a]	2.6	8.9	(36.4)	(122.1)	(159.2)

[a]Excluding military.
SOURCE: Bureau of Economic Analysis, U.S. International Transactions.

forth huge transfers of assets from one country to another. The U.S. dollar rose 34 percent against the currencies of its major trading partners between 1983 and 1985, then plummeted by 42 percent to recent lows principally because of fiscal and monetary—not trade or management—decisions (see Figure 2) (*The Economist*, 1988).

Thus, comparative costs for producing or sourcing in particular locations have often become more a function of exchange rates than of productivity or competitive managerial decisions. This argues for new geographical plant-

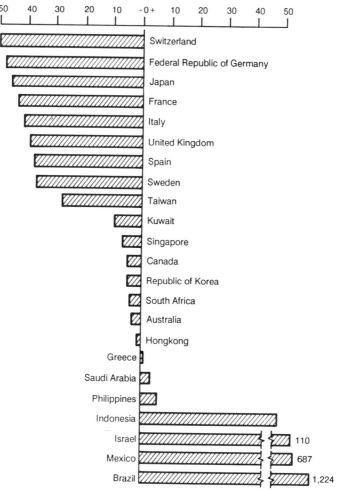

FIGURE 2   Exchange rate fluctuations. Percent change in the dollar (February 26, 1985–October 12, 1987) against national currencies. SOURCE: *The Economist* (1987a).

deployment tactics, market portfolios, and levels of organizational and lo-gistics flexibility not commonly found in past U.S. manufacturing strategies. Recently, the lower cost U.S. dollar has undoubtedly encouraged more man-ufacturing by Japanese firms in the United States. If the European Economic Community integrates in 1992 and maintains internal exchange rate parities, it could become a most attractive location for long-term investment, to the detriment of the United States. The entire power relationship between man-ufacturing and services groups is also changing profoundly in other areas.

• For example, many manufacturers now find that their medical care or insurance outlays for employees are higher than their own profits. Hence, new strategies creating "coalitions" with providers and insurers have emerged as key elements in cost control.[3] Deregulation has created more powerful transportation companies with intermodal han-dling capabilities that improve shipping efficiencies enormously, but also increase these carriers' bargaining power against manufacturers and other shippers (Cook, 1987). In addition, large money center banks now offer information, instant capital access, and worldwide connection advantages that few manufacturers can duplicate internally for managing their financial assets.

All "services" or "products" are really just means for providing satis-faction to customers. Thus, the boundary between services and manufactures is very fluid and varies widely over time. It is imperative that U.S. domestic economic and international trade policies recognize the extreme fluidity, substitutability, and mutual interrelatedness between manufacturing and ser-vices.

GOVERNMENT POLICY ISSUES

Recognizing these important facts as context, what are the central policy initiatives needed to support the effective use of technology in services? A basic presumption behind sound policy in the United States is that a com-petitive marketplace will allocate resources optimally among producing sec-tors, markets, and technologies. In most instances, market forces are quite sufficient in services. Government intervention would seem justified only in those situations where market imperfections or externalities make it unlikely that private initiatives can meet the challenge.

Within this general framework, an attempt has been made to identify a few major points where government interventions would have high leverage. Five basic policy directions would appear to be most productive:

1. macroeconomic and tax policies focused on improved capital formation rates, lower cost of capital, and lengthened time horizons for return on investment;

2. increased and better-targeted national investments in both hard and soft infrastructures supporting services;

3. restructured regulatory practices to improve the efficiency and innovativeness of the services sector;

4. a refocus on employment and human resources development policies more appropriate to the mobility and intellectual skills required for a services-dominated society; and

5. stronger recognition and exploitation of services-manufacturing interface potentials in international trade measurements and in trade negotiations.

Macroeconomic and Tax Policies: Capital Formation and Technology Investment

The basic macroeconomic and tax policies to support a strong services sector are quite compatible with those desirable for a healthy manufacturing sector—not surprising when one understands the substitutability and interrelatedness of the two sectors. High among the policies that would be most productive are measures to enhance capital formation and encourage long-term investment in research and technology.

In the past, national growth in productivity has tended to correlate highly with national capital formation rates and investments in technology and producing assets (Barras, 1986; Kendrick, this volume). Because of the high capital intensiveness in services (Quinn, this volume), this should be equally true for larger companies in both services and manufacturing. Capital formation is stimulated by policies that selectively encourage savings and investment over consumption. Decreasing taxes on earned interest or the double taxation on dividends would, of course, assist such a reemphasis, as would increasing the relative percentage of government tax revenues from consumption taxes—as opposed to those based on income production (Landau and Hatsopoulos, 1986). Although the economic benefits and political costs of these measures have been debated extensively, the substantial leverage of the benefits for and from the services sector has rarely been considered.

Because of the long time frames they involve, the development and application of major new technologies are especially sensitive to high capital costs or tax laws that bias decisions toward short-term investments. U.S. capital costs have been estimated to be significantly higher than those in Japan (Hatsopoulos et al., 1988). A simple calculation will show that a company investing $1 million at the recent U.S. average cost of capital (approximately 15 percent) can only afford to wait 4.7 years to recover an expected income of $500,000, whereas a Japanese company investing at its 6 percent capital cost can wait for 11.9 years. This goes a long way toward explaining the more "patient" Japanese outlook. The cycle for invention, development, and successful implementation of technological innovations in services is quite comparable to that in manufacturing: 3–5 years is a typical time frame for most major innovations to be effectively incorporated in production processes or to breach a new market as a "services" product (Guile and Quinn, 1988).

Radical innovations tend to take significantly longer than this. Federal Express's COSMOS II tracking system took 8 years, Citicorp's development of automated teller machine systems spanned more than 10 years, and the cellular mobile telephone waited 11 years from AT&T's public filing to market implementation. Currently, the U.S. tax law is among the industrial world's least attractive for long-term investors. Only those of Australia and Britain are worse (see Table 3). Return to a 6-month holding period for capital gains status would be of little benefit in stimulating needed long-term investments, but a minimum holding period of at least 3 years (with a significantly graduated tax reduction for longer-term gains) could help encourage both start-up enterprises and longer time horizons for corporate shareholders, whose short-term outlook is a constant complaint of corporate managers. Any capital gains tax reduction should, of course, be related to expected rates of inflation.

The electronic integration of world capital markets will make a national policy of arbitrarily low capital costs much more difficult to maintain, even for Japan. Nevertheless, selectively lowering capital costs for longer term investments (through something like a 3-year capital gains benefit) could be a viable strategy, especially if combined with less inflationary fiscal policies. Most notable among these would be decreasing national deficits to lower pressures on available capital sources as well as to ameliorate investors' uncertainties and inflationary expectations. Lower and more predictable cap-

TABLE 3   Capital Gains Taxes on Share Investments,[a] 1987

	Maximum Short-Term Rate (%)	Maximum Long-Term Rate (%)	Long-Term Qualifying Period	Annual Tax-Free Allowance (dollars)
Australia	50.25	50.25	1 year	—
West Germany	56	Exempt	6 months	543
Sweden	45	18	2 years	—
Britain	30	30	—	10,679
United States	38.50	28	6 months	—
Canada	17.51	17.51	—	22,650
France	16	16	—	44,336
Belgium	Exempt	Exempt	—	—
Italy	Exempt	Exempt	—	—
Japan	Exempt	Exempt	—	—
Holland	Exempt	Exempt	—	—

[a]Rates exclude local taxes.
SOURCE: Arthur Andersen.

ital costs would tend to lengthen investment horizons significantly. Increased long-term investment is essential to heightened technological innovation.

In addition, increased and better managed government investments in both hard and soft services infrastructures—research, education, communications, data collection, transportation, waste disposal, and health care systems— can improve the competitiveness of U.S. services and goods producers as much as many private investments could. Recent studies suggest high economic benefits from increased public investment in such infrastructures (Aschauer, 1987; *The Economist*, December 1987e). Historically, government-supported services infrastructure investments—starting with mail, road, canal, and water transport systems, and expanding later into higher education, basic research, electric power distribution, and health care systems—have been major components in building both U.S. productivity and human capital bases for a higher per capita income. Unfortunately, however, real public sector investments and capital stock per public worker have been allowed to drop steadily during the 1970s and early 1980s (see Figure 3).

A catch-up program is clearly called for, but the developing power of services technologies may now allow private concerns with their (presumed) efficiencies to provide certain public services more effectively than government bodies. A careful and constant reappraisal of such potentials versus the coordination, scale economies, or subsidies that only government investment could bring, is certainly warranted.

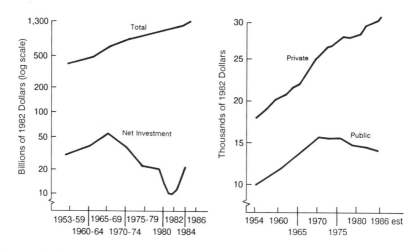

FIGURE 3   Real public sector investments (annual public spending averages in 1982 dollars) and capital stock per public worker (in 1982 dollars). Net investment and public capital stock data do not include defense expenditures. SOURCE: *The Economist* (1987d).

Services and Regulation: Improving Efficiency and Innovation

Services technologies have significantly restructured the nature of competition within and among industries. More explicitly, services technologies have made possible much wider ranges and more complex sets of direct and cross-industry competition. For a bank, a financial services "competitor" in one situation may also be a customer, supplier, co-venturer, or intermediary in another. Similarly, airlines have now become competitors, customers, and suppliers for postal services, travel agents, communications services, banks, and even retail establishments. Consequently, it is becoming ever less effective to regulate companies or industries on an industry or institutional basis (i.e., regulating banks or airlines as individual industries). In some cases, more efficient market structures will allow competition to replace regulatory requirements. In others, new forms of intervention may be desirable to overcome new market imperfections or to accelerate the development and deployment of emerging services technologies.

Increasingly, regulation is likely to be more equitable at a transactional level (i.e., establishing similar disclosure, safety, or insurance requirements for a particular transaction) regardless of which institution actually handles it, or at the functional level (i.e., establishing similar health, maintenance, environmental, safety, or electronics interface standards across all industries that might compete). Regulation at the institutional level (such as the Glass–Steagall Act) merely encourages disintermediation, i.e., bypassing of the overregulated institution. It also promotes inefficiencies by eliminating potential competitors and their alternative products. In addition, as Kendrick (this volume) mentions, it is imperative that regulations be goal oriented, rather than specifying means, and that these goals constantly be coordinated and reevaluated in a systematic manner across agencies.

Although intellectual property rights are not covered elsewhere in this volume, it is worth noting that better techniques for protecting intellectual property would be highly desirable. In industries providing services such as software, music and video recording, and data bases, intellectual property protection both domestically and across national borders is crucial. Information achieves its maximum value in a society when it is shared. Clearly understood and dependable patent structures—or enforceable copyrights—would encourage sharing of knowledge through more complete disclosure (including the patent itself), thereby allowing others to build more effectively on the protected concept. Currently, patent law lacks clear or reliable rules for defining infringement versus legitimate copying of software. Consequently, important software is often available only in noncopyable form, and computer companies expend considerable effort to prevent others from cloning and selling software they spent years and tens of millions of dollars to create. Table 4 shows the status of software protection in selected countries

TABLE 4   Status of Software Protection in Selected Countries: 1986

Country	Copyright	Trade Secret Unfair Comp.	Trademark	Patent
Brazil	Unclear[a]	Very Limited	—	No
France	Yes[b]	Unclear	—	Perhaps
Federal Republic of Germany	Yes[b]	Yes	—	No[c]
Indonesia	No	—	—	—
Japan	Yes[b]	Limited	—	Sometimes
South Korea	[d]	Limited	Yes	Sometimes
Taiwan	Yes[b]	—	Pending	—
United Kingdom	Yes[b]	Yes	Yes	Sometimes
United States	Yes[b]	Yes	Yes	Sometimes

[a]Legislation proposed.
[b]Special legislation passed or decree issued.
[c]Perhaps, if included in process.
[d]Legislation pending.
SOURCE: Nusbaumer, 1987b, pp. 216–217.

and suggests that international coordination is badly needed. Lack of adequate protection mitigates against innovation and wastes the valuable resources of innovators in seeking tricks to keep others from stealing and exploiting their efforts.

Another complex problem is posed by vertical integrations through information networks that enable retailers, distributors, manufacturers, and suppliers to tightly coordinate their efforts. While lowering costs for consumers and increasing U.S. competitiveness, such systems also contain potentials for anticompetitive behavior (*Business Month,* 1987; *Wall Street Journal,* 1987a). Developing appropriate competition-inducing guidelines for antitrust and fair practices interpretations will be a significant challenge; yet it could have a high payoff in competitiveness and also, as Kendrick states, in terms of productivity. By encouraging distribution efficiencies through restructuring and such technological developments, even Britain's somewhat archaic distribution system has improved to the point where many Japanese-made products can be delivered to U.K. customers at lower prices than those charged in Japanese retail outlets (*The Economist,* 1987c).

A final and very important aspect of regulation in services has to do with financial markets, economic risk, and control. Technological advances in communications and computing have increased the scale and speed of transactions and the interdependence of world economies in ways that can have profoundly favorable—but also potentially sudden and disastrous—consequences. The debate is just beginning on how to prevent large-scale international interventions or computer-based trading systems from overwhelming securities markets in the short run.

The Presidential Task Force on Market Mechanisms (the Brady Commission), appointed by President Reagan after the market crash of October 19, 1987, suggested oversight of all markets by one competent body, a more unified clearing system, consistent margins requirements for all major players, and expanded emergency techniques to slow or halt trading so that all players have more equitable access to the system (Presidential Task Force on Market Mechanisms, 1988). Other recommendations have focused on institutional corrections, such as changing specialists' roles and rules (*The Economist*, 1987b). Private groups such as the stock exchanges have often not needed government intervention to implement many of these reforms in their own self-interest, but others may require international coordination through government agreements.

The Bank for International Settlements is also setting forth proposals for banks in the Group of Ten industrial countries to have identical, well-defined, minimum standards for both core and secondary reserves. Such standards would mean that banks compete internationally under similar rules and are less likely to drastically stress the world financial system with costly, cascading failures (*The Economist*, 1987d). Yet they could also seriously affect the capacity of less developed countries to obtain needed loans. More attention to the interests of less developed countries and the implications for their international trade balances will be a core policy issue to ensure a more stable and continuously growing world economy.

Many maintain that the present complex of insurance mechanisms, hedging instruments, and open markets is sufficient to avert real disasters and that those who invest should bear their own risks (Norton, 1988). Government regulations in anticipation of unknown events always carry with them inefficiencies and risks. These must be weighed against potentials of massive losses for the United States or the entire world economy if matters go awry. This probably argues for some combination of increasing the technological capacities of major markets to handle peak trading loads and some carefully constituted mechanisms for limiting the nation's total economic risk, not complete nonregulation, however efficient that might seem in the short run. However, any proposals to selectively regulate U.S. markets should recall that in an integrated world marketplace, any inefficiencies introduced by U.S. regulation will quickly cause customers to seek more efficient solutions in the markets of other nations.

Human Resource Policies for a Services-Dominated Economy

As in other major policy areas, sound national human resources policies are not substantially affected by the national mix of services and manufacturing industries. At a specific level, worker adjustment programs are an important ongoing activity in any dynamic economy, and the requirements for good basic skills training for new labor force entrants are high but not substantially changed by recent technical advances or growth of services industries. A recent comprehensive review of the impacts of technology on employment (Cyert and Mowery, 1987, p. 169) concludes

New technologies by themselves are not likely to change the level of job related skills required for the labor force as a whole. We do not project a uniform upgrading or downgrading of job skill requirements in the U.S. economy as a result of technological change. This does not deny the need, however, for continued investment and improvement in job related skills of the U.S. work force to support the rapid adoption of new technologies that will contribute to U.S. competitiveness.

One important concern is a more sophisticated treatment of the issue of services wages in public policy debates. Comparing average wages in manufacturing versus those in the services sector is particularly misguided. Because great variations exist among specific job categories and industries, a more relevant focal point would be on the wage levels and opportunities in particular services industries or occupations versus those in individual manufacturing categories. Even then certain important adjustments should be made.

Today, services—not blue-collar jobs—provide the most available entry point for new, secondary, or part-time workers, as well as for high school students, females, and retirees entering or reentering the work force (see Table 5). When analyzing the wages in services, one should adjust for these entry conditions, the greater convenience and better working conditions, the lower experience requirements, and the more flexible hours offered by many services jobs. Such jobs are essential for multiple-earner families and for developing the attitudes, skills, and disciplines needed for more permanent job holding (Levy, 1987).

The safety net provided by widely distributed services jobs has undoubtedly helped by absorbing displacements from manufacturing into local jobs in the services sector. The economy has benefited if the wage levels in growing services areas—and the value added which is presumably necessary to support them—are higher than those that displaced blue-collar workers would have had to accept to keep their companies from going overseas. Further, as services company outputs become more substitutable for manufacturing tasks (notably in design, quality assurance, maintenance, accounting, marketing, or other support tasks), wages should progressively equilibrate between the two sectors, as data indicate they have begun to over the past five years.

Disparaging services wages in policy conversations seems singularly unproductive. Instead, an emphasis on retraining, adapting, and cushioning the personal costs for those displaced because of the loss of certain manufacturing jobs or because of changes in occupational mix would be more appropriate. At the most fundamental level, a more literate and numerate work force will

TABLE 5 Secondary Workers in the Labor Supply, 1980 Census Data

Industry	Total Employed (numbers in thousands)	Work Less than 35 Hours per Week (%)	19 Years Old and Under (%)	Female (%)
All manufacturing	21,194	9.1	4.3	31.9
Selected services				
Food and bakery stores	2,502	38.3	22.8	45.8
Gasoline service stations	627	25.8	22.5	16.7
Apparel and accessory stores	896	40.2	17.7	69.7
Eating and drinking places	4,181	49.8	30.8	59.6
Drug stores	490	39.2	20.2	60.6
Other retail trade	2,217	34.0	10.0	54.6
Private households	701	58.6	9.7	90.9
Hotels and lodging	1,052	29.6	9.7	62.3
Laundry, cleaning, and garment services	399	30.3	7.8	58.9
Entertainment and recreation services	1,007	39.9	16.6	40.4

SOURCE: O'Neill (1987, p.22).

be more adaptive to changing economic conditions. Therefore, basic education remains a high priority, especially if the basic education offered can prepare people for lifetime learning in connection with constantly changing job demands. In his chapter in this volume Kendrick also concludes that it is most important to stress education appropriate for services jobs, especially vocational training and in such programs as the Job Training Partnership Act.

Many studies have emphasized that education and human skills development are the foundations of value-added in many industries and provisions for lifelong education are critical for individuals to obtain and upgrade the skills called for in today's rapidly restructuring economy. Whereas job-specific training by companies is often the most appropriate type of training for economic adjustment, there are situations in which the benefits an employer can capture are significantly less than the social benefits of an enhanced skill base. In those cases public support of training and education is crucial. One particularly important concern is that between 20 and 30 percent of displaced workers with job experience lack basic skills (Cyert and Mowery, 1987). These individuals, with less formal education, are also the least likely to derive benefit from on-the-job training provided by employers and should be a primary target for public training and retraining initiatives. This priority is reinforced by Kutscher's observation (this volume) that job growth in the near future is likely to be mostly in those jobs currently held by those with higher levels of educational attainment.

Finally, at the college and university level, there are a host of initiatives that could strengthen the technology base of the nation with as much benefit to services industries as to manufacturing. The 1985 Report of the President's Commission on Industrial Competitiveness recommended greater funding for engineering education and expansion of the National Science Foundation's engineering research centers. The increasing technological intensity of services industries argues for sustained support of engineering education as much as does the need to keep U.S. manufacturing technologically dynamic.

Sectoral Risks and Diversification

Another key policy issue is the extent to which government should attempt to lower the country's economic and political risks by maintaining sectoral diversification between the goods and services sectors. In an increasingly global economy, and one in which the United States is more of an equal participant than an overwhelmingly dominant force, national security concerns with the industrial base may require special interventions. Critical issues include whether the growth of services threatens U.S. capabilities to maintain (1) a sufficient presence in international trade, (2) its desired military or defense posture, and (3) its strategic flexibility to obtain vital resources abroad. The latter two points may require special interventions by consciously diversifying national strategic alliances, stockpiling critical materials, or directly subsidizing specialized defense needs. The cost of dealing with these issues through general economic supports for various manufacturing industries would be extraordinarily high; asking consumers to subsidize noncompetitive industries for the sole purpose of maintaining a production surge capacity for certain defense contingencies is a very inefficient and wasteful way of maintaining our defense industrial base.

In the realm of commercial trade, services already account for a high percentage of U.S. and other developed countries' exports (see Figure 4). Yet there is general agreement that these figures are vastly understated. Data concerning services trade measure only a few categories of services activities. Many services are often embodied, but not separately accounted for, in product or technology transfers across borders. However, the most interesting point is generally overlooked: trade figures measure the benefits of products versus services sold in trade in quite different ways. When a product, say a mined raw material, is sold, the selling country registers the wholesale price of the product as its gain or inflow. Yet all the resources embodied in the product are lost to that country for further use. The real net gain on the sale is actually the profit on the sale, perhaps one-tenth of the sale price. On the other hand, when a services company completes a transaction abroad (e.g.,

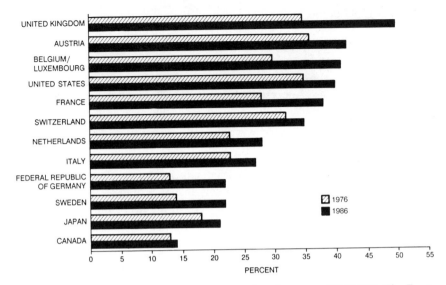

FIGURE 4   Exports of services as percentage of total exports. SOURCE: *The Economist* (1987b).

a banking deal or a bridge design in situ), only the net profit or fee is credited to the selling country. Yet no nonrenewable resources have left the country, and those who handled the transaction (and return to their parent country) may actually have increased value as human assets. Thus, international trade accounting conventions probably overstate the relative net value of products versus services sold in trade by at least an order of magnitude.

Some have suggested that it may be impossible for a country to establish meaningful competitive or comparative advantages in services (Deardorff, 1985). Given the capital intensity, economies of scale, and scope of some of the major services industries, some companies and industries can demonstrably achieve international competitive advantages in cost, quality, or flexibility. The more interesting question is comparative advantage. Because traded services depend so much on costs in the receiving country, do the relative factor endowments of the parent country really determine what is exported? Also, if a country had a factor advantage that favored particular services industries, could that shift trade patterns dangerously (Nusbaumer, 1987a)? Wright and Pauli (this volume) warn that the Japanese may be using such a factor advantage (availability of inexpensive money in Japan) in penetrating U.S. financial services markets. Because exploitation of this advantage will require substantial Japanese investments in their customers' countries (like the United States), there are grounds for negotiating fair access to Japanese services markets in return.

Despite U.S. strengths in many other services areas, there are significant foreign barriers to trade. Gönenç, in his chapter, identifies three other factors that may limit the growth of trade in marketed services: (1) the existence of culturally bound services demands, (2) the complexity and cost of internationalization in services businesses, and (3) the lack of incentive to engage in international trade or operations on the part of the many privately held, smaller services establishments. And in fact, the measured world market share of manufactured goods produced by U.S.-based companies has hovered around the 20 percent level since the 1960s, and manufacturing has enjoyed a strong recent export resurgence (*Fortune*, 1987). Given these and the high degree of interactive support and mutual substitution possible between services and manufacturing, it seems unlikely that U.S. trade as currently measured will shift dramatically toward services in the near future.

However, it is appropriate to focus more attention on the $700 billion international services trade, which is virtually all outside GATT (Shelp, 1987). Because so much of the services sector of most nations is typically government owned or managed, it will take years for a general agreement such as GATT to substantially ameliorate each country's unique constraints. Specific bilateral or limited multilateral negotiations seem the most likely road to progress in the near future. Open access for foreigners to the huge U.S. services markets will normally be a desirable quid pro quo in such negotiations. Federal Express has often been hampered in pursuing Fred Smith's vision of a global just-in-time economy (Smith, this volume) by Japanese government protectionist actions. An effective transportation and logistics system linking U.S. production and distribution units with worldwide sources will demand extensive modification of outdated world air cargo regulations and treaties, but the benefits of less regulation could be great for all parties.

The threat of tariff barriers is rarely effective in services trade, but selectively being able to deny takeovers of U.S. companies by foreign enterprises whose countries do not offer reciprocal access may be a valuable bargaining tool in obtaining "level-playing-field" conditions with our major competitors. The international airlines industry provides an interesting and timely case in point. Successive administrations awarded foreign airlines access to the richest U.S. travel cities, while settling for access to fewer cities abroad and heavy restrictions on U.S. airline operations there (*Wall Street Journal*, 1987b). Foreign airlines are now asking for cabotage—the right to fly domestic U.S. routes—and permission to own some U.S. airlines, so a possibility for redress appears to exist.

Services trade and merchandise trade are inseparable in many instances. Gönenç (this volume) points out the way in which Japanese services exports in transportation derive naturally from Japan's large merchandise export activity. Leverage can be brought against countries with strong product trade balances to allow U.S. manufacturing companies to operate under equivalent rules, including provision of U.S. services in support of their product sales in those countries. To do otherwise is to allow competitor nations to exploit their advantages in manufacturing, while denying the United States the right to benefit from its relative strengths in services.

Because most of the employment and benefits from providing services in international trade tend to be captured in the receiving country—with the exception of some services industries such as water transport or communications—there is serious doubt that direct services trade can make up the enormous product trade deficits which the United States has incurred recently. In the near future, services are most likely to help U.S. trade by lowering U.S. infrastructure costs and increasing the value added in exported manufactures through mechanisms described above. This area, along with aggressive efforts to break barriers to services trade and to U.S. services investments abroad, is where much of our strategic focus should be placed.

CONCLUSIONS

Policy discussions have generally derided, misunderstood, or underemphasized the role of services in the U.S. economy. This volume has attempted to systematically address certain important policy areas, by using the most up-to-date data available. These include the issues of productivity, investment, employment, regulation, and trade in services. This chapter attempts to integrate these analyses and set forth some explicit recommendations addressed to those major policy communities that can most dramatically enhance the effective development and use of technology in services. Although market forces are generally reasonably efficient in allocating research resources among services-producing sectors, markets, and technologies, a major reordering and refocusing of national priorities and attitudes toward the following could pay high dividends:

● Developing a strong services sector is an extremely desirable goal for the United States, which will support and enhance U.S. manufacturing competitiveness nationally and internationally. Manufacturing and services effectiveness are so intertwined that any attempt to enhance one at the expense of the other will be actively counterproductive.

● Macroeconomic and tax policies should focus on improving capital formation rates, lowering capital costs, increasing investments versus current consumption, and lengthening investment time horizons. Crucial, and feasible, elements include fiscal constraints on government expenditures to lower anticipated inflation rates and pressure on capital funds available, institution of a 3-year minimum holding period (and perhaps progressively higher incentives for longer term holdings) to obtain significant capital gains tax advantages, and a shift in government expenditures toward infrastructure investments including education and training and other "intangible" investments in social infrastructures.

● Because services technologies have created a much wider and more complex range of direct and cross-industry competition, new regulatory approaches and institutions are needed which intervene at a transactional or functional level, rather than on an industry or institutional basis (i.e., regulating sales transactions or safety functions, rather than banks or airlines as individual industries). Other new modes of intervention may be needed to protect proprietary knowledge and to enforce or enhance competitiveness in the new integrated information environments and cooperative structures permitted by modern services technologies.

• Human resources policies should be redirected to foster the mobility and intellectual skills required for a services-dominated society. Basic educational, job training, and wage policies should reflect the fact that services, not blue-collar, jobs now provide the primary entry point for new workers, secondary workers, minorities, and other less advantaged people entering or reentering the work force. A particular focus for public initiatives should be continuing education, including on-the-job-training for those with less than average formal education.

• Further attention to, and leveraging of, the strong U.S. services position in international trade should become a central element in U.S. trade policy. Intense pressures can be brought on selected foreign countries to break down barriers to services trade (and even product trade) by refusing to allow their companies to consummate takeovers of U.S. services enterprises or have access to desirable U.S. services infrastructures such as our major air routes or airports. Pressures should be brought to create level-playing-field access not just for product trade but also especially for the services accompanying that trade, where the United States is likely to have distinct competitive advantages. To do otherwise is to allow other nations to exploit their advantages in manufacturing, while denying the United States the benefit of its relative strengths in services.

Although there is no single new "program" or identifiable R&D "agenda" that calls for a special government action on behalf of services, the above recommendations do provide a catalogue of priority shifts that can make an enormous difference in the effective development of this most crucial sector of economic activity. With proper policy support for technology and trade development, services can continue to be the main engine for U.S. growth in jobs, GNP, and value added through the end of the century. Government's main task is to nurture the attitudes, infrastructures, and skills development that will make this possible.

ACKNOWLEDGMENTS

The authors gratefully acknowledge the important contributions of Dr. Jordan J. Baruch and Penny C. Paquette in developing this chapter, as well as the generosity of Bankers Trust Company, Bell & Howell, The Royal Bank of Canada, Braxton Associates, Bell Atlanticom Co., and American Express in supporting this research.

NOTES

1. Unfortunately, national economic data bases (as they are now compiled) are of limited use in understanding the services economy (Quinn, 1987). There are many problems of defi-

nition. Categories for collecting data still reflect the dominance of manufacturing concerns in an earlier period. Domestic categories do not capture many new services areas (such as software or health spas) until they become more mature industries. Also, data are not refined enough to sort out critical relationships. Step by step, as opportunities present themselves, new categories reflecting services activities should be substituted for the less relevant manufacturing details now collected.

Data showing services transactions between industries and in international trade are particularly weak. U.S. input-output tables are not available until 5–10 years after data collection. Various responsible federal groups have formed task forces to improve data and methodologies for analyzing productivity, wage, trade, unemployment, and output data for services. However, they have been hampered both by budgetary constraints and by the cost and difficulty of getting businesses to report in more detail.

2. Bell and Kettell (1983, p. 3) estimate that 95 percent of the daily volume in foreign exchange markets in 1983 was not direct commercial business but trading between the foreign exchange dealers of the world's international banks.

3. At the national level, the Washington Business Group on Health has been attempting to coordinate provider, payer, and government agency groups that need to cooperate on this issue and has started publishing *Business and Health* to bring the views of both private and public authorities to the fore.

REFERENCES

Aschauer, D. A. 1987. Is the public capital stock too low? Federal Reserve Bank of Chicago Bulletin (October).

Barras, R. 1986. A comparison of embodied technical change in services and manufacturing industry. Applied Economics 18(September):941–958.

Bell, S., and B. Kettell. 1983. Foreign Exchange Handbook. Westport, Conn.: Quorum Books.

Business Month. 1987. Taking control in the rag trade. (April):48–49.

Cook, J. 1987. If it isn't profitable, don't do it. Forbes (November 30):54–56.

Cyert, R. M., and D. C. Mowery, eds. 1987. Technology and Employment. Washington, D.C.: National Academy Press.

Deardorff, A. 1985. Comparative Advantage and International Trade and Investment in Services. Philadelphia: Fishman-Davidson Center, The Wharton School, University of Pennsylvania.

The Economist. 1987a. Economic and financial indicators. (October 17):130.

The Economist. 1987b. Taking stock. (October 31):14.

The Economist. 1987c. Purveyor to all nations. (December 5):16.

The Economist. 1987d. Bank regulation: Levelling? (December 12):92.

The Economist. 1987e. The wrong kind of squeeze? (December 26):32.

The Economist. 1988. Get ready for the phoenix. (January 9):9–10.

Faulhaber, G., E. Noam, and R. Tasley, eds. 1986. Services in Transition: The Impact of Information Technology on the Service Sector. Cambridge, Mass.: Ballinger Publishing Co.

Fortune. 1987. Crawling out of the trade tunnel. (December 21):43–44.

Guile, B. R., and J. B. Quinn. 1988. Managing Innovation: Cases from the Services Industries. Washington, D.C.: National Academy Press.

Hatsopoulos, G. N., P. R. Krugman, and L. H. Summers. 1988. U.S. competitiveness: Beyond the trade deficit. Science (July 15):299–307.

International Trade Commission.1982. The Relationship of Exports in Selected U.S. Services Industries to U.S. Merchandise Exports. Washington, D.C. (September).

Landau, R., and G. N. Hatsopoulos. 1986. Capital formation in the United States and Japan.

Pp. 583–606 in The Positive-Sum Strategy: Harnessing Technology for Economic Growth, R. Landau and N. Rosenberg, eds. Washington, D.C.: National Academy Press.

Levy, F. 1987. Changes in the distribution of American family incomes, 1947–84. Science 238(May 22):923–927.

Mark, J. 1982. Measuring productivity in the services sector. Monthly Labor Review (June):3–8.

Mark J. 1986. Problems encountered in measuring single and multifactor productivity. Monthly Labor Review (December):3–11.

National Academy of Engineering. 1983. The long-term impact of technology on employment. National Academy of Engineering Symposium, Washington, D.C. (June 30).

Norton, R. 1988. The battle over stock market reform. Fortune (February 1):18–26.

Nusbaumer, J. 1987a. The Services Economy: Lever to Growth. Amsterdam: Kluwer Academic Publications.

Nusbaumer, J. 1987b. Services in the Global Market. Amsterdam: Kluwer Academic Publications.

Office of the U.S. Trade Representative. 1983. U.S. National Study on Trade in Services. Washington, D.C. (December).

O'Neill, D. 1987. We're not losing our industrial base. Challenge (September-October):19–25.

Presidential Task Force on Market Mechanisms. 1988. Report of the Presidential Task Force on Market Mechanisms. Washington, D.C.: U.S. Government Printing Office.

Quinn, J. B. 1977. National policies for science and technology. Research Management (November):11–18.

Quinn, J. B., 1983. Overview of current status of U.S. manufacturing. Pp. 8–52 in U.S. Leadership in Manufacturing. Washington, D.C.: National Academy Press.

Quinn, J. B. 1987. The impacts of technology in the services sector. Pp. 119–159 in Technology and Global Industry, B. R. Guile and H. Brooks, eds. Washington, D.C.: National Academy Press.

Quinn, J. B., and B. R. Guile. 1988. Managing innovation in services. Pp. 1–8 in Managing Innovation: Cases from the Services Industries. Washington, D.C.: National Academy Press.

Sauvant, K. 1986. International Transactions in Services: The Politics of Transborder Data Flows. New York: Westview Press.

Shelp, R. 1986. Understanding a new economy. Wall Street Journal (December 23):20.

Shelp, R. 1987. The folly of excluding the services from the trade framework. International Management (November):104.

Vollmann, T. 1986. The effect of zero inventories on cost (just in time). Pp. 141–164 in Cost Accounting for the '90s: The Challenge of Technological Change. Montvale, N. J.: National Association of Accountants.

Wall Street Journal. 1987a. Computers as marketing tools. (March 18):1.

Wall Street Journal. 1987b. Freeing up the international skies. (December 1):36.

Advisory Committee on Technology in Services Industries

Chairman
JAMES BRIAN QUINN, William and Josephine Buchanan Professor of Management, Amos Tuck School of Business Administration, Dartmouth College

Members
DONALD N. FREY, Professor of Industrial Engineering and Management Sciences, Northwestern University, and former Chairman of the Board, Bell & Howell Company

PAUL F. GLASER, Chairman, Corporate Technology Committee, Citicorp

THOMAS R. KUESEL, Chairman of the Board, Parsons, Brinkerhoff, Quade & Douglas, Inc.

FREDERICK W. SMITH, Chairman and Chief Executive Officer, Federal Express Corporation

ERIC E. SUMNER, Vice President, Operations Systems and Network Planning, AT&T Bell Laboratories

Contributors

THOMAS L. DOORLEY is managing partner and founder of Braxton Associates. The thrust of his work has been toward strategy management, that is, working to achieve integration between strategy definition and effective implementation. He has led the extension of Braxton's strategy management concepts and client base into consumer and service organizations. Before founding Braxton, he was a senior consultant on business strategy and organizational issues and a business unit manager at Arthur D. Little, Inc. Mr. Doorley holds an M.B.A. from Columbia University and a B.S. in chemical engineering from Pennsylvania State University.

FAYE DUCHIN has been studying the economic and social implications of technological change for a decade. She is director of the Institute for Economic Analysis and Research Professor at the Graduate School of Public Administration at New York University. She is the author (with Wassily Leontief) of *Military Spending: Facts and Figures, Worldwide Implications, and Future Outlook* and *The Future Impact of Automation on Workers*. Professor Duchin received a B.A. in psychology from Cornell University and a Ph.D. in computer science from the University of California, Berkeley.

RAUF GÖNENC is at present an administrator with the Organization for Economic Cooperation and Development (OECD). Having received his Ph.D. in international economics and finance at Paris University, he has since been responsible for various OECD projects on the growth of high-technology services areas and has written reports on the emergence of the software industry, the funding of information technology sectors, and the structural changes in financial and health services. His chapter was written while the author was on leave from the OECD to a large Australian services organization.

BRUCE R. GUILE is associate director of the National Academy of Engineering Program Office. Before joining the Academy in 1984, Dr. Guile worked as research associate with the Berkeley Roundtable on the International Economy and as a management consultant. Dr. Guile holds a bachelor's degree in English literature and computer science from Heidelberg College, a master's of public policy from the University of Michigan, and a Ph.D. in public policy from the University of California, Berkeley.

JOHN W. KENDRICK has been professor of economics at the George Washington University since 1956. Earlier, he was a member of the senior research staff at the National Bureau of Economic Research (1953–1956) and an economist with the U.S. Department of Commerce (1946–1953) to which he returned as chief economist in 1976–1977. He has authored a dozen books on national income, wealth, and productivity.

RONALD E. KUTSCHER is associate commissioner for economic growth and employment projections of the U.S. Bureau of Labor Statistics (BLS). He directs the BLS program that develops medium-term projections of the U.S. economy, covering gross national product, industry output and productivity, and employment by industry and occupation. He has an extensive background in developing projections of changes in an economy, particularly as they affect the labor market.

JEROME A. MARK is associate commissioner for productivity and technology of the U.S. Bureau of Labor Statistics (BLS). He is responsible for development of the U.S. government measures of productivity for the economy and its component sectors and industries. He is also responsible for the BLS research on productivity measurement and its studies of the factors affecting productivity.

GUNTER PAULI is founder and chief executive officer of the European Service Industries Forum, a network of leading services companies in Europe. He is also president of PPA & Co., a Belgian-based international consulting firm.

JAMES BRIAN QUINN is the William and Josephine Buchanan Professor of Management at the Amos Tuck School of Business Administration at Dartmouth College. Dr. Quinn earned a B.S. from Yale, an M.B.A. from Harvard, and a Ph.D. from Columbia University; he joined the Tuck faculty in 1957. Professor Quinn is an authority in the fields of strategic planning, the management of technological change, and entrepreneurial innovations. He has held fellowships from the Sloan Foundation, the Ford Foundation, and the Fulbright Exchange Program. In addition to consulting with leading U.S. and foreign companies and publishing extensively on corporate policy issues, Dr. Quinn has the distinction of recently being named the dean of a Japanese business school.

STEPHEN S. ROACH is a principal and senior economist of Morgan Stanley & Co., Inc., and has overall responsibility for the firm's forecasting and analysis of economic activity in the United States. His recent published work has covered a broad range of topics from technology and foreign trade to the consumer and capital spending. Mr. Roach holds a Ph.D. in economics from the New York University and a bachelor's degree from the University of Wisconsin. Before joining Morgan Stanley in 1982, he was vice president, economic analysis, for the Morgan Guaranty Trust Company. Prior to that he served for six years on the research staff of the Federal Reserve Board in Washington, D.C., where he supervised the regular preparation of the official staff projections of the U.S. economy. He has also been a research fellow at the Brookings Institution in Washington, D.C.

FREDERICK W. SMITH is chairman and president of Federal Express Corporation. Each weekday, the company completes more than 840,000 door-to-door deliveries of critical items in 40,000 communities throughout the United States, Canada, Puerto Rico, Europe, and Asia. The company operates a fleet of McDonnell Douglas DC10s, Boeing 727s, and Cessna 208s as well as more than 17,000 computer- and radio-dispatched vans. The work force has grown to more than 48,000 men and women. Mr. Smith formulated the idea of Federal Express while a student at Yale University. He graduated in 1966 and served as an officer in the U.S. Marine Corps. Mr. Smith serves on the boards of directors of General Mills, First Tennessee National Corporation, and ALSAC/St. Jude Children's Research Hospital.

RICHARD W. WRIGHT is Helen Simpson Jackson Distinguished Professor of International Management in the Geo. H. Atkinson Graduate School of Management, Willamette University, Oregon. He is the author of seven books on international management and finance.

Index